British Nuclear Weapons and the Test Ban

This book provides an overview of how the UK tried to maintain and modernise its strategic and tactical nuclear weapons during 1974–1982, whilst also pursuing a comprehensive nuclear test ban treaty.

The core question addressed in the book is how a test ban treaty would impact the reliability and safety of the UK's nuclear weapons and how this would constrain and limit efforts to secure a comprehensive treaty that would prohibit nuclear testing indefinitely. An added complication lay in the fact that a ban treaty could also prevent or limit the UK's ability to test new warhead designs to replace its existing tactical nuclear weapons and a new strategic successor system to Polaris. How all of this played out between 1974 and 1982, when the UK announced its decision to acquire Trident and the US decided that a test ban treaty was no longer in its security interests, is discussed. A detailed review, based on the available materials in the UK National Archives, also looks at the aims and objectives of UK nuclear tests in Nevada and on the decisions taken on the Chevaline warhead and its Trident replacement. The book also considers whether there was a far greater threat to the UK nuclear programme from shortages of skilled craftsmen and industrial action at the Atomic Weapons Establishment at Aldermaston than from a test ban treaty. It also looks at whether nuclear defence trumped arms control objectives during this period.

This book will be of much interest to students of British politics, nuclear proliferation and Cold War History.

John R. Walker is Senior Associate Fellow, Royal United Services Institute and at the European Leadership Network, and Senior Research Fellow at the Department of Science and Technology Studies, University College London. He worked in the Arms Control and Disarmament Research Unit, Foreign and Commonwealth Office, from 1985 to 2020, serving as its Head from 2014 to 2020.

Cold War History

Series Editors:
Michael Cox
London School of Economics & Political Science, UK
Odd Arne Westad
John F. Kennedy School of Government, USA

In the new history of the Cold War that has been forming since 1989, many of the established truths about the international conflict that shaped the latter half of the twentieth century have come up for revision. The present series is an attempt to make available interpretations and materials that will help further the development of this new history, and it will concentrate in particular on publishing expositions of key historical issues and critical surveys of newly available sources.

Secrecy, Public Relations and the British Nuclear Debate
How the UK Government Learned to Talk about the Bomb, 1970–83
Daniel Salisbury

US Foreign Policy and the End of the Cold War in Africa
A Bridge between Global Conflict and the New World Order, 1988–1994
Flavia Gasbarri

European Socialist Regimes' Fateful Engagement with the West
National Strategies in the Long 1970s
Edited by Angela Romano and Federico Romero

NATO and the Strategic Defence Initiative
A Transatlantic History of the Star Wars Programme
Edited by Luc-André Brunet

Technological Innovation, Globalization and the Cold War
A Transnational History
Edited by Peter Svik and Wolfgang Mueller

British Nuclear Weapons and the Test Ban
Squaring the Circle of Defence and Arms Control, 1974–1982
John R. Walker

For more information about this series, please visit: www.routledge.com/Cold-War-History/book-series/SE0220

British Nuclear Weapons and the Test Ban

Squaring the Circle of Defence
and Arms Control, 1974–1982

John R. Walker

Routledge
Taylor & Francis Group

LONDON AND NEW YORK

First published 2023
by Routledge
4 Park Square, Milton Park, Abingdon, Oxon OX14 4RN

and by Routledge
605 Third Avenue, New York, NY 10158

Routledge is an imprint of the Taylor & Francis Group, an informa business

© 2023 John R. Walker

British Library Cataloguing-in-Publication Data
A catalogue record for this book is available from the British Library

Library of Congress Cataloging-in-Publication Data
Names: Walker, John R., 1960– author.
Title: British nuclear weapons and the test ban : squaring the circle of defence and arms control, 1974–82 / John R. Walker.
Description: Abingdon, Oxon; New York NY : Routledge, 2023. | Series: Cold war history | Includes bibliographical references and index.
Identifiers: LCCN 2022055448 (print) | LCCN 2022055449 (ebook) | ISBN 9781032451633 (hardback) | ISBN 9781032451640 (paperback) | ISBN 9781003375708 (ebook)
Subjects: LCSH: Nuclear weapons—Government policy—Great Britain—History. | Nuclear arms control—Great Britain—History. | Treaty Banning Nuclear Weapon Tests in the Atmosphere, in Outer Space and Under Water (1963 August 5)
Classification: LCC U264.5.G7 W353 2023 (print) | LCC U264.5.G7 (ebook) | DDC 355.8/251190941—dc23/eng/20221115
LC record available at https://lccn.loc.gov/2022055448
LC ebook record available at https://lccn.loc.gov/2022055449

ISBN: 978-1-032-45163-3 (hbk)
ISBN: 978-1-032-45164-0 (pbk)
ISBN: 978-1-003-37570-8 (ebk)

DOI: 10.4324/9781003375708

Typeset in Bembo
by Apex CoVantage, LLC

Contents

Foreword

With the war in Ukraine refocusing attention on the existential risk represented by nuclear weapons, an understanding of the diplomacy, political decisions and technological challenges that have fashioned the current nuclear landscape is essential. Political progress is rarely linear, and some of the more crabwise involutions of international diplomacy that culminated in the Comprehensive Test Ban Treaty in 1996 are described in this book with an immediacy and depth that should cause us to reflect upon how further progress can be made at this, another critical juncture in global nuclear weapons policy.

Since my time in government, I have sought to advance the causes of non-proliferation and transparency around nuclear weapons. Last year, the UK Government made a surprise announcement, lifting the cap on its overall nuclear weapons stockpile and ceasing to publish data relating to its stockpile and missile numbers. In addition, the review appeared to expand the number of scenarios in which nuclear weapons would be used – a reversal of the unspoken, decades-old commitment to a 'sole purpose' policy – that nuclear weapons are solely for the purpose of deterring attacks on the UK or its allies. It has been suggested that these announcements (delayed owing to COVID) were designed to harmonise with the Trump Administration's more hawkish approach to nuclear weapons, but is markedly out of step with President Biden's ambition to reduce the salience of nuclear weapons in foreign policy, rather than increasing it.

Fifteen years ago, as the United Kingdom's Secretary of State for Defence, I was responsible for overseeing the beginning of the replacement of the Trident nuclear-armed submarines. This first-hand experience of being involved in ensuring congruence between nuclear defence and foreign policy has shown me the critical nature of the missteps in the Integrated Review, a verdict only sharpened by the nuclear sabre-rattling from Russia as their conventional military capabilities in Ukraine undergo further degradation. During John Walker's 35 years in the Arms Control and Disarmament Research Unit at the Foreign Office, the UK cemented its reputation as one of the most forward-leaning of nuclear weapons states in reducing its nuclear weapons stockpile (and being transparent about that) as well as advancing the disarmament element of Article VI of the NPT. At a time when we are dissipating that valuable goodwill amongst our allies, and compromising our record of leadership in non-proliferation, this work reminds us of how we built that reputation in the first place – and perhaps indicates how it might be regained.

Since I left the Ministry of Defence in 2008, we have seen a shift in the international security landscape. Debates around proportionality, complicated by cyber warfare and an increasing tendency for hostile powers to operate in the 'grey zone', that designedly inchoate margin between mere hostility and an act of war, means that some have advanced the argument that this means a greater role for nuclear weapons. These arguments have only been amplified by Putin's attempts at nuclear blackmail. To me, these calls for a greater role for nuclear weapons show a (potentially) fatal complacency, both about the effects of the employment of nuclear weapons, but also the risk of a cyber attack on command and control systems and the consequent inadvertent employment of these weapons. No system is invulnerable, especially in a world where we rely more and more on complex computer systems that fewer and fewer of us actually understand.

With all this in mind, John Walker's compelling book is not only a valuable contribution to the scholarship of nuclear policy, but extremely timely. In this very readable analysis, he describes the competing priorities of nuclear safety and the transatlantic relationship during a crucial period of the Cold War, helps the reader see the mechanics of the inner workings of government as policy is devised and implemented and describes how consensus began to form around the desirability and practicability of a Comprehensive Test Ban Treaty – which would eventually bear fruit in 1996. As he explains, the British state faced the twin challenges of modernising its nuclear deterrent while simultaneously working towards a ban on the testing of nuclear weapons – and he shows us the means by which successive governments sought to square that circle in concert with the US and USSR. These twin imperatives continue to preoccupy policymakers today, and Dr Walker's account offers real lessons and precedents that can be applied, valuably, in the current climate.

Dr Walker's 35 years of experience in the Arms Control and Disarmament Research Unit at the Foreign Office including, latterly, 5 years as its head – means he is uniquely placed to write this scholarly analysis of the UK's approach to arms control and nuclear testing between 1974 and 1982. Though the chronological focus is narrow, designedly focusing on a hitherto under-researched period, many of the tensions and trade-offs described will resonate with policymakers today.

I have had the pleasure of knowing, and occasionally working with John over a period of around five years. I have witnessed his leadership in the non-proliferation, disarmament and verification space and know that many senior officials and ministers have relied upon and profited from his wise counsel. This book is a valuable contribution to the history of the post-war nuclear deterrent in the UK and should cause us to reflect on what we can all do to advance the cause of non-proliferation.

by Lord Browne of Ladyton
October 2022

Preface

When I was interviewed for the post of Senior Research Officer in the Foreign and Commonwealth Office's Arms Control and Disarmament Research Unit in January 1985, I was asked how I might defend the UK's then refusal to participate in negotiations for a Comprehensive Nuclear Test Ban Treaty (CTBT). I replied along the lines that a treaty could not be reliably verified given the state of remote detection technologies, adding that I didn't necessarily agree with such an assessment. That answer must have been sufficient as the following week the FCO offered me the post, which was for three years with a possibility of extension for a further two. I ended up staying for just over 35 years, eventually heading the Unit and retiring in May 2020. After about 1987, I no longer focused on CTBT issues, working instead primarily on chemical and biological weapons arms control, but my intertest in CTBT issues had been piqued and I never lost my interest. In 2004 an opportunity presented itself for a two-year part-time secondment to a Southampton University British nuclear weapons history project under Professor John Simpson funded by an Arts and Humanities Research Council grant. The original plan was that I would write the arms control aspects chapters in the planned three volumes covering the years 1952–1958, 1959–1964 and 1965–1972. By this time I was back dealing with CTBT issues in the FCO as the UK's representative on the Preparatory Commission's Working Group B dealing with on-site inspection issues. Once I had started to rummage around the documents in The National Archives at Kew, it quickly became clear that there were masses of open papers on the UK weapons programme and its relationship to a CTBT, the vast majority of which had not been used before in academic books or scholarly articles. Professor Simpson and I agreed that it would now be a better idea to write all this up in a separate book, hence *British Nuclear Weapons and the Test Ban 1954–1973 Britain, the United States, Weapons Policies and Nuclear Testing: Tension and Contradictions*, published in 2010. It was always my intention to write a follow-up as more papers on testing issues became available at Kew, but realised that would likely have to wait until I retired from the FCO. Since then, when the opportunity presented itself, I collected copies of relevant papers from Kew against the day that I would have the time to write what is now this present volume. I cannot be the only author who has had his/her plans delayed by the COVID pandemic, but once The National Archives opened up again, I was able to crack on and complete the manuscript.

There have been many scholars and officials over many years that have directly or indirectly maintained my interest in British nuclear weapons policies and programmes, and I am most grateful for their advice and help: John Ainslie, Lorna Arnold, Jonathan Aylen, Andrew Barlow, Professor John Baylis, Brian Burnell, Dr Geoffrey Chapman, Bruce Cleghorn, Dr Alwyn Davies, Denis Fakley, Mike Fazakerley, Professor Eric Grove, Ken Johnston, Professor Matthew Jones, Ian Kenyon, Glyn Libberton, Peter Marshall, Dr Richard Moore, Dr Frank Panton, Dr Joe Pilat, Dr Kate Pyne, Sir Clive Rose, Peter Sankey, Professor John Simpson, Dr Kristan Stoddard, David Summerhayes, Robin Woolven, and all the speakers and participants at the Charterhouse British nuclear history conferences over the years.

Many thanks are due to the three external reviewers who recommended to Routledge that they should publish this book; they made many helpful comments and suggestions, which I have tried to reflect and incorporate in the text wherever possible. Hopefully, this makes for a better book.

It goes without saying that none of these esteemed folks, some of whom are now sadly no longer with us, have any responsibility whatsoever for the contents of this book, nor does it reflect FCO views. Any errors of fact or interpretation are mine alone.

I am also most grateful to Lord Browne of Ladyton for agreeing to write a short foreword for this book.

The staff at Routledge – Andrew Humphrys, Devon Harvey and Apex Covantage – Sathyasri Kalyanasundaran have all been first rate and helped ensure a smooth production process.

And finally, but by no means least, thanks to my partner Lorna Miller for all her patient understanding when I head back north to Glasgow for Celtic home games. This book is for you.

1   Introduction

During the Cold War, Britain had an ambivalent attitude to nuclear weapons arms control and disarmament. On the one hand, British Prime Ministers, Foreign and Defence Secretaries and their officials in 10 Downing Street, the Foreign and Commonwealth Office and the Ministry of Defence wanted to see an end to the arms race with disarmament measures under effective international control, but on the other hand, they were acutely aware of the actual and potential adverse implications for the UK's own nuclear weapons programme arising from such measures. London generally much preferred to keep out of the bilateral US-USSR nuclear arms control negotiations since resulting treaties could prevent the modernisation of UK nuclear weapons. UK forces were tiny in comparison with the massive stockpiles in the American and Soviet arsenals. Britain's own nuclear deterrent was seen by both Labour and Conservative governments as an absolutely essential requirement for its own national security. As Margaret Gowing and many other scholars working on the history of the UK's nuclear weapons programme have noted over many years, independence and deterrence have been the distinguishing and recurring features of British nuclear policy.[1] In doing so, however, they have not addressed in detail British policy on a nuclear test ban treaty in particular and how it might impact the UK strategic deterrent and tactical nuclear weapons in the1950s, 1960s, 1970s and 1980s.[2] Maintaining this capability as technology and threats changed and evolved over the years of the Cold War provided continuing challenges for the UK scientists, engineers, civil servants and Ministers charged with responsibility for the nuclear weapons programme. One of the crucial areas where these challenges and arms control and disarmament policies came together were the negotiations for a Comprehensive Nuclear Test Ban Treaty (CTBT), which has been the Holy Grail of arms control since the 1950s; and one, moreover, inextricably linked to the prevention of nuclear proliferation since states could not acquire an effective and reliable nuclear weapons capability, especially one containing thermonuclear weapons, without testing. The essential dichotomy can be put succinctly: support for a CTBT was important in order to control and end the nuclear arms race, but a test ban could also hobble the UK nuclear deterrent by preventing the development of new warheads for replacement strategic delivery systems and potentially undermining the safety and reliability of Britain's small

DOI: 10.4324/9781003375708-1

stockpile too. Such competing policy imperatives were particularly acute in 1957 and 1958 as the UK sought thermonuclear weapons whilst also feeling the growing international and domestic pressures calling for a halt in atmospheric nuclear testing.[3] Once the UK had successfully tested a thermonuclear weapon design in November 1957, which helped to pave the way for restored UK–US nuclear cooperation in July 1958, unbridled British support for a CTBT became possible. And in this respect, Prime Minister Harold Macmillan played a critical role. No subsequent Prime Minister took as supportive and active interest in achieving a test ban as Macmillan, though James Callaghan came close as we shall see here. Although the efforts to secure a comprehensive test ban prohibiting nuclear tests in all environments proved elusive in the early 1960s, primarily because of Soviet opposition to on-site inspection – a sine qua non for any such treaty for the British and Americans, the UK, US and USSR eventually signed a Partial Test Ban Treaty on 5 August 1963. (It entered into force on 10 October 1963.) However, this treaty limited testing to underground only, and thereafter the US and USSR lost all interest in a comprehensive nuclear test ban treaty. Attention moved instead to negotiating a Nuclear Non-Proliferation Treaty, which was opened for signature on 1 July 1968 and which nevertheless had been a British foreign policy priority for Harold Wilson's Labour Government and to which it had made important contributions. Thereafter, the US and USSR focused on bilateral strategic nuclear arms control from 1969 – constraining anti-ballistic missile defence systems and strategic launchers for ballistic missiles. They duly signed an Anti-Ballistic Missile Treaty on 26 May 1972 and agreed on an interim limitation on strategic launchers (SALT 1) – also in 1972. Discussions continued in Geneva at the multilateral disarmament bodies – first the Eighteen Nation Disarmament Committee (ENDC) and then its successor, the Conference of the Committee on Disarmament (CCD) on test ban issues, but these were fruitless.[4] Although the US and USSR agreed on two treaties with some constraints on nuclear testing, the Threshold Test Ban Treaty (TTBT), which limited the yield of tests to 150 kilotons, in 1974 and Peaceful Nuclear Explosion Treaty (PNET) in 1976, the superpowers evinced no further interest in negotiating a CTBT until President Carter made it, and more ambitious strategic arms reductions, priorities for his new Administration in 1977.[5] These treaties did not enter into force until 11 December 1990, in part driven by US anxieties over the need for effective verification measures. Moscow was receptive to Carter's initiative. Initial bilateral discussions on the scope and content of a CTBT began on 13 June 1977.[6] Britain felt that it ought to participate in any negotiation on a test ban, so the Labour government led by Prime Minister James Callaghan asked the US for its agreement. This was forthcoming and the Soviet Union readily agreed too. To be seen supporting such a key nuclear arms control measure, which was also considered crucial to sustaining support for the NPT, came naturally to a Labour government. There was strong back bench opposition to nuclear weapons within the Labour Party; and even in the Cabinet, there were complaints about the 1974 UK underground nuclear test in Nevada, the first since 1965.[7] In addition, the timing of test ban negotiations and their successful conclusion came at an inopportune time for the UK's own nuclear weapons programme. Testing was still required to finalise the design of the new warhead for Chevaline – the improved front end for the Royal Navy's Polaris missiles.[8] Officials and Ministers believed that Chevaline was essential for the continued credibility of the UK's deterrent in the face of probable Soviet ABM developments and deployments in the coming years. Looking further ahead, the Atomic Weapons Research Establishment (AWRE) and the Ministry of Defence (MOD) had plans for tests of new nuclear warhead designs that would be relevant for a future strategic successor system for Polaris should Ministers so decide, and for the potential modernisation of the UK's tactical nuclear weapons that consisted of WE177 gravity bombs in service with the Royal Navy and Royal Air Force. The first of these tactical weapons, the high-yield WE177 B, had been deployed from September 1966 with the low-yield WE177As entering service between 1969 and 1972, and the intermediate-yield WE177Cs between 1973 and 1977. WE177's original design life was 12 years,[9] so decisions on a replacement would soon fall due. History was, therefore, repeating itself as once again as the UK found itself torn between support for a CTBT in order to halt and reverse the nuclear arms race and simultaneously trying to maintain its own minimum nuclear weapons capability. How and whether the UK managed to square this particular circle is the principal purpose of this study. British nuclear history scholars have thus far largely neglected this aspect of the weapons programme and its relationship to arms control and disarmament policies and the wider context of the Anglo-American special relationship more generally. There is also an interesting internal dynamic here between competing policy imperatives represented by AWRE, the MOD, the Foreign and Commonwealth Office, the Cabinet Office and 10 Downing Street as the principal players in this saga. All British nuclear tests had to be approved by the Prime Minister, and the final authority to fire also lay with the PM. The Foreign Secretary was usually informed too, but he had no veto. Given that the UK's only place to test its nuclear devices was at the Nevada Test Site (NTS) in the US following decisions taken in 1962 and the absence of anywhere else suitable, US Presidential approval was also required for UK tests. Furthermore, given the centrality of the 1958 UK-US Agreement for the Cooperation on the Uses of Atomic Energy for Mutual Defence Purposes (the UK-US Mutual Defence Agreement – MDA) to the British nuclear weapons programme in terms of the information, special nuclear materials, equipment and components made available to the UK, any discussion of British policies would unavoidably also have to include the US dimension and the vagaries of the US interagency system of policy-making, especially when the agencies were bitterly divided on fundamental policy questions as they would be on a test ban. This is also true in the context of the eventual tripartite negotiations on a test ban treaty with the USSR as the UK probably spent as much, if not more, time discussing test ban issues with the Americans as it did formally with the Russians in Geneva. Topics of fundamental centrality to this story, such as nuclear weapons stockpile reliability and safety and the role of national seismic stations (NSS), cannot be told or understood without reference to the US position or positions, which limited the universe of potential policy choices and outcomes open to the British.

There is a vast body of primary sources available in The National Archives on Britain and its nuclear weapons programmes covering the years 1974 to 1982. These years also have strong echoes from the 1954–1958 era when the UK also had to grapple with pressures to cease nuclear testing and develop its own thermonuclear weapons before a moratorium on nuclear testing came into force. Documentation on the tripartite negotiations in Foreign and Commonwealth Office departmental files in The National Archives (TNA), Kew is essentially complete with no significant omissions. Britain's role in these negotiations and the positions it adopted on such key topics as verification and treaty duration will be discussed in Chapter 2. This will not be a blow-by-blow account of the treaty text drafting process in each of the negotiating rounds – such accounts of complex and lengthy treaty negotiations can be very dry, tedious (and often were too for the participants) and not necessarily informative to the general reader, so instead the focus will be on the main themes, associated problems and on the eventual outcome in 1982. TNA files do offer a very detailed account of the negotiations, including the numerous UK and US bilateral discussions that are a major feature of the story, so it would be easy enough for anyone interested to follow up on any of the issues recorded in the countless submissions, minutes, letters, telegrams, meeting records and policy papers.

Even though there is a significant body of source material (see the Bibliography for a list of The National Archive files in the AIR, AVIA, CAB, DEFE, FCO and PREM classes) directly, or indirectly relevant to this study, there are still gaps in the open historical record where papers have still been retained under section 3 4 of the 1958 Public Records Act. In recent years the Nuclear Decommissioning Authority has recalled numerous files from Kew in the ES (Aldermaston) and AB (United Kingdom Atomic Energy Authority UKAEA) classes for review, something which further hampers the work of scholars working on British nuclear weapons history.[10] In many cases, this is for sound non-proliferation reasons where it would not do to reveal nuclear warhead design information – something that would be a breach of the Non-Proliferation Treaty and 1958 MDA, manufacturing secrets, or any vulnerabilities or capabilities of in-service weapons systems. This problem is particularly acute in the case of the scientific and technical objectives and outcomes of the UK's nuclear tests where there is little publicly available. However, we do have some detail, which enables the historian to assemble a general overview on the purpose of the ten tests conducted between May 1974 and April 1982. Unlike the atmospheric nuclear tests conducted in Australia, Malden Island and Christmas Island between 1952 and 1958, we only have the yield for four of the UK's underground tests conducted in the period covered by this book – all the rest are expressed in ranges. British nuclear tests and plans and where and how they seem to fit into modernisation plans and, most importantly, options for both future nuclear strategic and tactical weapons are considered in Chapter 3; the nuclear tests are also listed in chronological order in tabular form in the Annex along with as detailed a description of their purpose as is possible to derive from available papers in The National Archives.

As noted earlier, the core intention of this book is to explore how the issue of nuclear weapon stockpile reliability and safety would be affected by a test ban treaty. This goes to the heart of the matter and became the main obstacle to a truly comprehensive test ban. Data was needed from underground nuclear testing to inform and shape future weapon design requirements, but such data, experience and the confidence derived from testing were also required to ensure certification of reliable and safe warheads in existing stockpiles. Throughout the period covered by this book, stockpile reliability and safety became a pressing question for UK officials shaping policy and conducting the treaty negotiations in Geneva. This question became a problem precisely because the weapons laboratories in both the UK and the US assessed that an indefinite test ban treaty would be incompatible with the maintenance of a reliable and safe stockpile in the longer term (and not a particularly long term at that as we shall see). All of this, and the stockpile issue's impact on the negotiating position of the UK and the US in Geneva, and on this assessment's wider implications for UK non-proliferation policy are analysed in Chapter 4. This is a long, complex and convoluted saga, which is why Chapter 4 is unavoidably the longest.

In the 1970s, Britain was developing a new improved front end for its Polaris ballistic missiles, initially known as the Polaris Improvement Programme and later as Chevaline. However, Chevaline ran into delays and discussions in the MOD started to turn towards a possible need to extend the life of the existing Polaris ET317 nuclear warhead. Despite this, Chevaline was eventually completed at great expense and had overcome major engineering challenges. Chevaline was finally deployed in 1982 when HMS Resolution left the Clyde Submarine Base at Faslane on its first patrol armed with new warheads – two instead of three on the front-end. By the end of the 1970s, a decision became due on a Polaris replacement given the very long lead times from the point of a decision to acquire a new delivery system – new submarines, new missiles and warheads, not to mention the support infrastructure – to when it would be first deployed on operational service. Initial steps on preparing the ground for such a decision had been taken by James Callaghan (Labour Prime Minister) in 1978, but following the election of Conservative Prime Minister Margaret Thatcher in May 1979 there was never going to be any doubt that the UK would decide to replace Polaris given her firm views on the fundamental importance of nuclear deterrence for Britain's security. The question was with what. However, the most likely answer was the US Trident C4 submarine-launched ballistic missile rather than cruise missiles.[11] Whatever the choice, a new British warhead design would inevitably be needed and then tested.[12] The relationship of Britain's nuclear tests in the late 1970s to such a new warhead and the finalisation and production of the Chevaline design will be reviewed in Chapter 5. This chapter also discusses, albeit briefly, British thinking on replacements for its WE177 tactical weapons.

Chapter 6 offers some concluding remarks about Britain, nuclear weapons and the test ban treaty and whether Ministers and officials managed to square the circle between support for a treaty banning tests and simultaneously maintaining and modernising the UK's nuclear weapons capabilities. The conundrum was most clearly expressed in a proposed agreed UK line that could be put to the US Secretary of State when he met the Foreign Secretary Dr David Owen in August 1977. Officials proposed that the UK should say:

> A ban on nuclear testing would pose special problems for Britain in maintaining the effectiveness of our deterrent force, and in developing a successor system if the government should so decide in the future. We have nevertheless given our full political support to President Carter's initiative in getting negotiations going on a CTB, and we have wished to join in these. But both for the immediate improvement, and for the long term maintenance, of our nuclear deterrent we should be placed by a CTB or moratorium in a position where we should need to enjoy further cooperation from the US both in the field of nuclear warhead technology and in that of delivery systems.[13]

Arguably to some extent the problem facing the British as summarised in this line was solved by the actions of the US and USSR. The Reagan Administration's decision to suspend the test ban talks, and its later determination that a test ban was no longer in US interests coupled with the Soviet Union's technically absurd insistence that the UK must host ten seismic stations to monitor compliance with a three-year Treaty removed any chance that a treaty would be concluded in the sort of time frame that could constrain, or otherwise damage UK nuclear weapon capabilities. An interesting question is whether a truly comprehensive nuclear test ban treaty was ever really likely to be a practicable proposition in the period 1977–1980. Hopefully, the answer to this question will become evident as the chapters of this story unfold. The overall structure and content of this book, therefore, aim to describe and explain British nuclear weapons policies and their relationship to the British nuclear test ban treaty policy in a largely chronological narrative. Given that these topics have not been discussed in the academic literature to date as noted earlier, it is important to lay out the basic story and thereby open the way for other scholars of British nuclear weapons history to look at this period, its meaning and its implications in greater analytical detail.

However, before proceeding to the narrative, we should pause here to recall why nuclear weapons states conducted nuclear tests in the first place. This sets a key contextual starting point for this study. A senior UK Ministry of Defence (MOD) official, Denis Fakley, who played a key role in this story, explained to the Geneva Conference of the Committee on Disarmament (CCD) in April 1976 that nuclear weapon states tested for four main reasons. The first of these concerned the maintenance of stockpiled weapons in an entirely safe yet serviceable condition; the second was for the investigation of the effects which nuclear explosions have on other items of equipment; the third was to enable the modification of nuclear explosive devices so that they could be fitted into new delivery systems; and, finally to enable the study of the physical principles of nuclear devices.[14] On the issues of stockpile maintenance, Fakley went on to say that like other types of hardware, nuclear weapons aged, but they tended to age more rapidly than other weapons because of the characteristics of the special materials used in their construction. From time to time they had to be repaired, something which frequently required the replacement of some components. These replacements were quite likely to be different from the originals since available materials will have changed in the interim and because of new stricter safety and security regulations. Such changes were usually small, but whether or not they were significant could not be ascertained without testing. Fakley also stated that in order to ensure that refurbished weapons remained safe and serviceable, it was necessary to conduct a full-scale test.[15] However, a reduced scale test could offer the necessary reassurances in some circumstances.

Fakley's statements on stockpile safety and reliability strike at the main theme of this book. Furthermore, the date of his remarks is also particularly significant, as will hopefully become apparent in the subsequent chapters, since he made it to an international audience more than a full year before the UK, US and USSR tripartite nuclear test ban treaty negotiations began in Geneva on 13 July 1977.

Notes

1 See, for example, Margaret Gowing, *Independence and Deterrence Britain and Atomic Energy 1945–1952, Volume 1: Policy Making & Volume 2: Policy Execution*, Basingstoke, MacMillan, 1974; Lawrence Freedman, *Britain and Nuclear Weapons*, London, Macmillan, 1980; G.M. Dillon, *Dependence and Deterrence: Success and Civility in the Anglo-American Special Nuclear Relationship, 1962–1982*, Aldershot, Dartmouth Publishing, 1983; John Simpson, *The Independent Nuclear State, Britain, the United States and the Military Atom*, Basingstoke, Macmillan, 1984; Peter Malone, *The British Nuclear Deterrent*, London, Routledge, 1984; Ian Clark and Nicholas J. Wheeler, *The British Origins of Nuclear Strategy 1945–1955*, Oxford, Oxford University Press, 1989; John Baylis, *Ambiguity and Deterrence: British Nuclear Strategy 1945–1965*, Oxford, Oxford University Press, 1995; Peter Hennessy, *Cabinets and the Bomb*, Oxford, Oxford University Press/British Academy, 2007; Richard Moore, *Nuclear Illusion, Nuclear Reality: Britain, the United States and Nuclear Weapons, 1958–64*, London, Palgrave, 2010; John Baylis and Kristan Stoddard, *The British Nuclear Experience: The Roles of Beliefs, Culture and Identify*, Oxford, Oxford University Press, 2014; Tanya Ogilvie-White, *On Nuclear Deterrence: The Correspondence of Sir Michael Quinlan*, London, Adelphi Book, 421, The International Institute for Strategic Studies, June 2014; James Gill, *Britain and the Bomb: Nuclear Diplomacy, 1964–1970*, Stanford, Stanford University Press, 2014; Matthew Jones. *The Official History of the UK Strategic Nuclear Deterrent, Volume I: From the V-Bomber Era to the Arrival of Polaris, 1954–1964*, Abingdon, Routledge, 2017, and *Volume II: The Labour Government and the Polaris Programme, 1964–1970*, Abingdon, Routledge, 2017. John Baylis and Yoko Iwama, Joining the Non-Proliferation Treaty Deterrence, Non-Proliferation and the American Alliance, Abingdon, Routledge, 2020, Chapter 3 John Baylis, 'Britain, the Deterrence/Non-Proliferation Dilemma and the NPT.'

2 Mark Hoffman, editor, *United Kingdom Arms Control Policy in the 1990s*, Manchester, Manchester University Press, 1990.

3 John R. Walker, *British Nuclear Weapons and the Test Ban 1954–1973: Britain, the United States, Weapons Policies and Nuclear Testing: Tension and Contradictions*, Farnham, Ashgate, 2010.

4 TNA FCO 66/875, Tripartite Consultations on a Comprehensive Test Ban Treaty (CTBT): Geneva, 13 July 1977 Draft Position Paper No. 11: Historical Background, 23 June 1977.
5 Matthew J. Ambrose, *The Control Agenda: A History of the Strategic Arms Limitation Talks*, Ithaca and London, Cornell University Press, 2018.
6 TNA FCO 66/874, Draft Report of U.S. Delegation on First Meeting of Bilateral U.S.-U.S.S.R. Discussions on a Comprehensive Test Ban, 13 June 1977. For the rationale for a test ban treaty see, for example, George Rathjens and Jack Ruina, 'Commentary on the New Test Ban Treaties', *International Security*, Vol. 1, No. 3, Winter 1977.
7 TNA CAB 128/54 CC (74) 21st Conclusions Minute 3, 27 June 1974 and TNA CAB 129/178, C (74) 85, Nuclear Testing, Note by the Prime Minister, 31 July 1974.
8 TNA DEFE 13/1478, British Nuclear Test Programme, V.H.B. Macklen, 19 April 1978.
9 John R. Walker, *A History of the United Kingdom's WE177 Nuclear Weapons Programme From Conception to Entry Into Service 1959–1980*, London, BASIC, March 2019.
10 Robert Booth, 'British Nuclear Archives Withdrawn Without Explanation', *The Guardian*, 23 December 2018.
11 Kristan Stoddard, *Facing Down the Soviet Union: Britain, the USA, NATO and Nuclear Weapons, 1976–1983*, London, Palgrave Macmillan, 2014, pp. 59–63. See also TNA DEFE 19/275, Duff-Mason Report, December 1978.
12 TNA DEFE 19/275, John Hunt to Prime Minister, Future of the British Deterrent, Part III: System Options, paragraph 18, 7 December 1978.
13 TNA 66/900, W.J.A. Wilberforce, Defence Department to Mr Moberly and Private Secretary, Secretary of State's Talks with Mr Vance: A Possible Quid-pro-quo from the Americans for our Renunciation of our Nuclear Test Programme, 9 August 1977.
14 Statement at an informal meeting of the CCD, CCD/492, pages 1–5, 21 April 1976.
15 This is an interesting statement, since as of 1976 (and subsequently) the UK itself had never conducted such a test.

2 The 1977–1982 Tripartite Test Ban Treaty Negotiations and Aftermath

Introduction

A comprehensive test ban treaty had long been the Holy Grail of nuclear arms control as explained in the previous chapter.[1] The first serious efforts to negotiate such a treaty began in 1958 between the US, UK and USSR. International anxieties over the adverse health consequences of radioactive fallout from the very large megaton range yield atmospheric tests conducted by the US and Soviet Union in the early and mid-1950s also provided a compelling driver for action to stem the threat posed by such tests to public health and the environment. However, these efforts in the Conference on the Discontinuance of Nuclear Weapon Tests ended in 1962 with the principal protagonists as far apart as ever on such key verification issues as the role of on-site inspection and the number of seismic stations to be hosted on the participants' national territory. Action then moved to the new multilateral Eighteen Nation Disarmament Committee, but that solved nothing and after the US, UK and USSR eventually agreed on a Partial Test Ban Treaty in the summer of 1963, the domestic and international pressures for a comprehensive test ban were much reduced. Superpower attention and efforts now moved to the negotiation of a Nuclear Non-Proliferation Treaty from 1965. A Non-Proliferation Treaty was also a key UK objective and the UK itself played a key role in these negotiations.[2] Both the US and USSR moved nuclear testing underground and evinced no further interest in a comprehensive test ban. Instead, anxieties over the development of anti-ballistic missile warheads and the development of multiple independently targetable re-entry vehicles (MIRVs) provided further incentives to test. From 1965 until 1977 the pace of nuclear testing by the superpowers increased dramatically. In that period, for example, the US conducted 386 tests with 56 tests held in a single year – 1968.[3] The Soviet Union conducted 296 tests.[4] In the same time frame, the UK conducted a mere five tests. There were, hardly surprisingly, no trilateral or multilateral test ban treaty negotiations in this period. However, the US did agree on a Threshold Test Ban Treaty with the USSR in 1974, which limited the yield to no more than 150 kilotons, and a separate agreement on a Peaceful Nuclear Explosions Treaty in 1976.[5] Non-nuclear weapons states never regarded these treaties as an adequate substitute for a CTBT, and in this respect, they were quite right.

DOI: 10.4324/9781003375708-2

When the prospect of a threshold ban treaty became apparent to the UK in March 1974,[6] FCO and MOD officials concluded that Ministers would certainly not want to disassociate themselves from such an initiative.[7] In terms of the potential impact of a threshold treaty on UK nuclear defence interests, the UK view was clear. In the absence of a verified test ban treaty, the UK needed to retain the freedom to conduct the tests required to maintain the credibility of the deterrent. As a general point, it was thus important that any agreement that the US might reach with the USSR should not interfere in any way with the UK's own current and planned nuclear tests.[8] As it happened, the UK was just about to resume testing after a self-imposed gap of nine years. The Fallon test took place on 23 May 1974 – see Chapter 3 and the Annex for further details. Fortunately for the UK, the US had provided assurances to the MOD that it was taking British interests fully into account in the development of the threshold test ban treaty. Dr Kissinger, moreover, had promised the British Ambassador in Washington that the US would keep the UK fully informed of the developments in the negotiations for a threshold treaty. MOD officials saw no reason why any threshold that the US would accept could cause the British nuclear programme any problems.[9] There was a feeling nevertheless in some quarters in London that the whole idea was a gimmick – an effort to give President Nixon something to agree to in his summit meeting with Soviet leader Leonid Brezhnev when he went to the Moscow summit in June.

A CTBT continued to be discussed in the Conference of the Committee on Disarmament (CCD) in Geneva in the 1970s, but the major obstacle to progress continued to be the technical difficulties of evolving an effective and acceptable system of verification.[10] In July 1973, an informal group of experts meeting in Geneva demonstrated that there was still a category of seismic events that the then-present capabilities of remote seismic detection (and discrimination between an earthquake and explosion) could not identify with certainty. It would thus be possible for a state to continue clandestine development of nuclear weapons by underground testing, which would escape detection. For the UK, this was the crux of the problem. So long as the USSR insisted that national verification was sufficient to guarantee compliance with a test ban treaty, it was difficult to see how an effective and necessary on-site inspection regime could be agreed upon. Research on lowering the threshold of detection was underway in some states such as Sweden, but not in the UK at that time[11] – a marked contrast to the pioneering role played by AWRE Blacknest in the early 1960s on teleseismic research.[12] British contributions to CTBT discussions had been based on its programme looking into the possibilities of remote seismic monitoring, particularly the problem of discrimination and between earthquakes and explosions and had presented a Working Paper (CCD 440) in the CCD in 1973 on the value of seismogram modelling as a source of discrimination between seismic sources.[13] Nevertheless, UK experts were of the view that despite progress in remote seismic detection, a would-be violator could still test at yields of 10 kilotons or below without fear of detection.[14] As we will see in Chapter 4, this figure of 10 kilotons would prove critical. Given such a background and the UK position, it is no surprise that London rejected a draft Soviet test ban treaty that was presented in the United Nations General Assembly in September 1975 as it contained no provisions for on-site inspection and would rely wholly on national means of verification, would exclude peaceful nuclear explosions from the treaty's scope and only enter into force when all nuclear weapons states joined it. Given Chinese and French opposition to a test ban, early entry into force was thus not at all likely. Although the UK had not demanded on-site inspection as the US had done, the UK had instead insisted on 'effective' or 'adequate' verification, which reflected the same assessment as the US – namely the inability of national means to detect and identify the source of all seismic events. Exactly what constituted 'effective' or 'adequate' verification was never defined. The acceptable level of risk here in a CTBT verification regime was (and is even today) a political decision. At that time neither the Russians nor Americans were willing to move off their positions – no on-site inspection on Soviet territory and on-site inspections as an essential element in any treaty, respectively. In these circumstances, both FCO and MOD officials believed that it was in the UK interest to maintain its traditional insistence on 'effective' verification since this would allow a degree of flexibility in responding to any change in Soviet or US positions. UK dependence on US assistance for its nuclear weapons programme, however, made it imperative that the UK should not get out of step with Washington on a CTBT, or on any aspect of nuclear disarmament. This would become a familiar refrain in the years ahead. Hence, the tentative UK idea for a possible annual quota of tests, first raised in the 1975 spring session of the CCD, went no further following an expression of US (and Soviet) opposition.[15] Although the Soviet Union revised its draft treaty in 1976 to include the possibility of voluntary on-site inspections, this was still not a sufficient movement for the UK or US to engage.[16]

President Carter and a new US policy

President Jimmy Carter had made progress in bilateral nuclear arms control a prominent part of his election campaign in 1976, and he was keen to move quickly on a test ban treaty and give it high priority in his foreign policy if elected.[17] In fact, Carter favoured ending all nuclear testing 'instantly and completely'.[18] Carter's very personal and moral stance provided the decisive impetus for new negotiations for a test ban.[19] A major US policy study had revealed that there were differences of opinion in Washington on the advantages of a test ban treaty, but that had been true in the 1950s and early 1960s. Subsequently, the US Secretary of State Cyrus Vance proposed on 28 March 1977 during his visit to Moscow that the US and USSR should begin negotiations on a comprehensive nuclear test ban treaty. The UK Labour government of the time was keen to take part too – not least because of its support for disarmament initiatives, but also because the UK's only nuclear test site was in Nevada and that any ban on US testing would prevent the UK from doing so too as there were no viable alternatives available.[20] Nevertheless, Prime Minister James Callaghan told President Carter that he could count on the full support of the UK on his test ban policy.[21] In short, politically the UK had to remain in favour of a CTBT despite having reservations about the military and security consequences of such a treaty.[22] As the Foreign and Commonwealth Office (FCO) observed, the US was certainly aware of UK views on the security implications of a test ban, and that probably a majority opinion amongst officials shared them.[23] The Foreign Secretary, David Owen, was concerned that the internal UK position implied a degree of caution about a CTBT, and that it risked underplaying the international benefits of a treaty – a view that did not endear him to Sir Herman Bondi, the MOD's Chief Scientific Adviser.[24] Dr Owen also believed that the UK and the US should work to mitigate the potential adverse consequences for nuclear defence, and certainly attached great importance to British participation in the negotiations.[25] Fred Mulley, Secretary of State for Defence, knew of the diverging views in Washington and certainly thought that the UK should not appear to take sides in Washington,[26] something that was going to be difficult to achieve in the years to come. This was by means a unique posture in UK–US relations since if the UK were to align itself too clearly with the view that lost the battle in the inter-agency process, that could well compromise British influence in future. By 6 May 1977 Prime Minister Jim Callaghan, Chancellor of the Exchequer Denis Healy, the Foreign Secretary and the Defence Secretary had all formally agreed that the UK should take part in the US/USSR negotiations from the outset.[27] Callaghan duly wrote to President Carter on 12 May to ask that the UK should be invited to take part.[28] Consequently, later that month the Americans and Russians readily accepted that the UK should indeed participate in a new round of tripartite negotiations in Geneva.[29] Participation in the CTBT negotiations was a departure from the UK's customary stance on nuclear arms control where it had resolutely remained aloof from bilateral US-USSR strategic nuclear arms control negotiations since British forces were but a tiny fraction of superpower stockpiles. However, there were two key issues in the SALT talks that did directly concern the UK: non-transfer and/ or non-circumvention clauses in a SALT treaty that would have prevented, or constrained the US from providing nuclear weapons design information, components and materials and delivery systems to the UK. London repeatedly badgered Washington throughout the 1970s on these points to ensure that the US did not agree to any Soviet proposals that would have resulted in treaty language detrimental to UK interests.[30] In this respect, the UK was very much an indirect participant in the bilateral US-Soviet negotiations, but in the case of a CTBT the UK had sought and been granted a front-row seat. This makes British participation in the tripartite test ban talks of particular intertest as it was the only location where the UK was directly engaged on nuclear arms control in this period and until the next time the CTBT negotiations were underway in the 1990s.

US and Soviet experts held preliminary discussions in Washington from 13 to 16 June 1977. The Americans kept the UK closely informed and supplied London with records of the meetings and copies of all the major statements.[31]

The first round of negotiations opened in Geneva on 13 July, concluding on 27 July.[32] So as in the late 1950s, the UK was once again running with the hare and hounds.[33] In short, working for a test ban treaty whilst simultaneously trying to maintain the UK's independent nuclear deterrent, which this time round would rely on underground testing to enable the UK to develop, test and deploy both future strategic and new tactical nuclear weapon designs and concepts to replace Polaris and the WE177s. This will be discussed in greater detail in Chapter 3. For the Foreign Secretary at least, Dr David Owen, whilst recognising the importance of defence interests, these were not more important than a CTBT – a view that would place him at odds with the MOD over the following two years.[34] Moreover, as we will see in Chapter 4, concerns over maintaining stockpile safety and reliability, especially in the US, would also further significantly hobble the UK's position in the Geneva negotiations. So how did this impact the positions taken by the UK in the negotiations? This chapter will discuss the main British negotiating objectives and the main elements of the draft treaty that started to emerge with a particular focus on the verification aspects given their centrality to a treaty's effectiveness and acceptability to the US and UK, which as in 1958–1962, not only shaped the treaty's content but also presented one of the principal obstacles to progress. Much the same would apply between 1977 and 1980.

The UK's objectives

London's main objectives, which were seemingly shared by the US, in negotiating a test ban treaty were to curb the qualitative development of nuclear weapons without adversely affecting Western security; and to help prevent their proliferation to other countries.[35] This latter objective required the sort of treaty that could attract the adherence of the non-nuclear weapons states, such as India and Pakistan, that had kept the nuclear weapons option open by not acceding to the Non-Proliferation Treaty. India had tested a so-called peaceful nuclear explosive device in May 1974. This point was particularly important to the UK since it now had disquieting intelligence information about the extent to which Pakistan was pressing ahead with its own nuclear weapons plans. At the outset of the negotiations in July 1977 the UK's negotiating position could be summarised as follows. First, the UK wanted to see a multilateral test ban treaty, which was adequately verified and included a ban on peaceful nuclear explosions. PNEs had to be banned since any test ban treaty that permitted such explosions would not be comprehensive in scope, would encourage non-nuclear weapons states to develop their own PNEs and could in due course be one factor liable to undermine Western security.[36] Second, the UK would regard a moratorium on nuclear testing as acceptable only if it were in binding form, of fixed duration and separate from a multilateral treaty.[37] The Prime Minister would go on to tell the House of Commons that 'a unilateral decision (to halt testing) by this country would not improve the prospects of achieving our objective of a permanent treaty banning all nuclear explosions'.[38] The UK recalled the Soviet breach of the 1958 testing moratorium in September 1961 and officials were worried that any new moratorium might end up becoming a de facto ban without any verification required under a legally binding treaty.[39] A key perception held by UK officials in 1977 was that the Soviet breach of the moratorium back in 1961 had given the USSR a temporary advantage.[40] Third, the timing of the start of any ban on tests would be an important consideration for the UK – the MOD's plans for a crucial nuclear test in 1978 were no doubt (see Chapter 3) uppermost in officials' minds. Fourth, the UK still had to decide how to minimise any security difficulties for its nuclear weapons programme that might arise from a test ban.[41] Indeed a test ban agreement should not jeopardise the freedom of the US and UK to exchange information on nuclear technology.[42] The Prime Minister had told officials that he would wish to know, before signing a CTBT, how the credibility of the UK deterrent could be maintained during a moratorium lasting, for example, five years.[43] As for verification, this would be a key issue in the negotiations and ensuring that a future treaty was adequately verified would, as it had been in 1958–1962, be a key British objective. Indeed, the UK's main objective in the negotiations was to seek agreement on some measure of on-site inspection, which would add significantly to the UK's ability to deter Soviet non-compliance with the treaty.[44] Even though the UK thought that Soviet acceptance of mandatory on-site inspection was unlikely, the Prime Minister nevertheless made clear at the outset that it should be a UK negotiating objective.[45] The purpose of an on-site inspection, at least in UK thinking, would be to investigate those seismic events that gave rise to a suspicion that a breach of the CTBT had occurred. In 1967 UK Ministers had decided that the UK would no longer require on-site inspections in a CTBT, but this policy was never made public, nor was it shared with the US, where Washington was still strongly in favour of on-site inspections and would react badly to any such UK change in policy. This was the main reason the decision was kept private. However, by the mid-1970s the UK technical assessment was, as already noted, that remote seismic means of detection and identification were not then sufficient to determine whether an event was a nuclear explosion, the only way to check would be in an on-site inspection.[46] In 1977 the UK view on the efficacy of seismic verification was that the detection level would be about 4.5 magnitude, which roughly corresponded to yields in the range of between 3 and 50 kilotons depending on the geology in which the test was conducted. Moreover, the UK did not see any real prospect of improving significantly upon this level of performance of remote seismic detection means.[47] Surprisingly, at least for the UK, the US had a more optimistic view, namely that the detection level would be down at about one to three kilotons and the identification level somewhat below ten kilotons.[48] All this was to have implications for the design of a verification network of seismic stations in a test ban treaty as well as the significance of the potential for clandestine cheating and an ability to derive a meaningful military advantage.

British experts in the MOD, FCO, and Cabinet Office and at AWRE were clearly aware of the potential disadvantages of a CTBT even before the negotiations began. Over a period of years, a treaty could adversely affect the credibility of the Western deterrent more rapidly than that of the USSR since the proficiency of UK and US warhead design teams would decline with scientists leaving the laboratories through lack of challenging work, but this would be less of an issue for the USSR given its ability to direct and control its experts in a command economy and closed society. Even under the best verification regime, the USSR could still conduct small-yield nuclear tests of significant military value without being detected.[49] In the longer term, a CTBT would seriously inhibit the development by the UK of improved nuclear weapon systems – a disadvantage only partially offset by access to existing US nuclear warhead designs. So in short, a CTBT was not without risk, but the political benefits in terms of preventing proliferation and slowing the arms race were on balance worth the risk.

In formal Whitehall terms, the Foreign and Commonwealth Office led the negotiations, but the UK position was carefully coordinated beforehand between the FCO, Cabinet Office and MOD. We should note here that, unlike the US where the Joint Chiefs of Staff (JCS) had a strong, discordant and dissenting voice, the British Chiefs of Staff had no direct role at all in UK CTBT policy-making. Policy was kept under constant review with coordination back in London under the Cabinet Office. The UK delegation was led by an Ambassador rank official (initially Percy Craddock until April 1978 and then John Edmonds until the end), with a deputy from the Ministry of Defence (Denis Fakley from Defence Science whom we met at the end of Chapter 1).[50]

The Geneva negotiations: the early stages 1977–1978

The UK/US and USSR entered the negotiations with widely differing opening positions.[51] Nevertheless, progress in the early rounds of the tripartite negotiations seemed to be encouraging to the British.[52] As the UK and US expected, the three main problems centred on the Soviet Union's traditional positions on a CTBT: that peaceful nuclear explosions (PNEs) should be permitted, that no treaty should not enter into force until both France and China were also on board, and that external means of verification (i.e. teleseismic and other national technical means such as satellites and remote detection of radionuclide debris) were sufficient to monitor a state's compliance with a test ban. On-site inspection would have to be voluntary and procedures would have to be worked out only *after* the state had accepted a request for an inspection.[53] The UK and the US correctly argued that there was no way to distinguish a PNE from a military test and as such prevent some military benefits arising from a PNE.[54] They also made the point clearly that it was unrealistic to expect either France or China to join a test ban treaty at that time given their well-known obdurate positions. However, the Soviet Union abandoned the first two of these requirements in November 1977 when Soviet President Leonid Brezhnev proposed that there should be a 'moratorium covering nuclear explosions for peaceful purposes along with a ban on all nuclear weapons tests for definite period'.[55] Brezhnev's concession was certainly a promising sign that suggested that perhaps rapid progress towards a treaty might be made after all. Despite the apparent Soviet concession, the Russians argued that PNEs were still important to the Soviet economy, but they were prepared to accept a moratorium of a fixed period of three years equal to the fixed duration that they were prepared to accept for a test ban treaty. The issue was thus being kicked down the road for resolution later. Both the UK and the US put a counter-proposal in which a test ban treaty and PNE protocol should be of indefinite duration, but with a provision that would allow the parties to withdraw after four or five years if other states continued to test. It was apparent to Percy Craddock after the first round of negotiations that the Soviet movement on the PNE issue was a prerequisite for progress in the negotiations as a whole.[56] At the same time as movement on the PNE question, the Soviet position on Chinese and French accession to the treaty simultaneously with the UK, US and USSR also changed – it was now no longer necessary. A further step forward came when both sides made significant concessions over the scope and content of the treaty's verification regime. The Prime Minister himself strongly expressed the view in early February 1978 that the negotiations should be wound up during the summer and asked that everything should be done to keep up their momentum, and the delegation in Geneva was instructed accordingly.[57]

On-site inspections

The UK's starting position in 1977 had been that there was no known remote method of determining unambiguously whether an underground event was man-made in origin, and, if so, was due to a nuclear explosion. Such conclusive evidence could only be obtained in an on-the-spot investigation into the presence of radioactive materials (a set of radionuclides such as Argon-37 and the ratios between them) that could only have been created by a nuclear explosion. When the fourth round of negotiations opened in January 1978, the US, which thus far had demanded mandatory on-site inspections after a suspicions seismic event that might have been a nuclear explosion, accepted a 'voluntary' system following UK arguments in favour of such a position, which the Americans had evidently found persuasive.[58] This might help explain why the FCO Minister of State Evan Luard told the House of Commons on 10 January 1978 that he believed that 'the negotiations can be brought to a successful conclusion'.[59] However, this came with the important proviso that in the case of the nuclear weapon states, certain rights and functions of inspection teams were to be agreed upon in advance. If they were not, there could be endless haggling over the procedures that could keep inspectors away from the area of interest for months, which was important because some of the observables created by an underground nuclear test, such as aftershocks and some radionuclides of intertest, were perishable.[60] Both the UK and the US wanted to see mandatory inspection rights and procedures so that effective inspections could be conducted in all three states to the same set of rights and obligations for inspection teams and inspected state parties. These rights would include inter alia the use of the best technical inspection equipment available. However, the Russians, whilst accepting the need to use a broad range of inspection equipment types, insisted that detailed discussion of lists and concepts for their use in an inspection would need to be deferred for negotiations at the Joint Consultative Commission (JCC) *after* entry into force of the treaty. It was all part of a pattern as the USSR always sought to defer discussion and agreement on as many parts as possible of the detailed intrusive verification provisions to the JCC, which would convene only after the treaty had entered into force. John Edmonds thought that satisfactory compromises on OSI issues could be agreed if the other parts of the treaty were also agreed.[61] However, it is worth noting that finalising the OSI provisions in the eventual Comprehensive Nuclear Test Ban Treaty that was concluded in 1996 was by no means easy; and, moreover, a further 25 years of effort in the Preparatory Commission's work on inspection equipment and concepts of use showed just how much scope there was for genuine technical differences over equipment lists and specifications.[62] The acceptance of voluntary on-site inspections would also mean that the state party being challenged would have the option of refusing a request for an on-site inspection, but as the US admitted to the UK privately, it was only really a concession to reality because a sovereign state could always refuse an inspection. In any case, a refusal would be tantamount to an admission of guilt, and other states could draw the appropriate conclusions and respond accordingly. In contrast, the USSR saw the US positions on on-site inspections and the installation of internal seismic stations on Soviet territory, particularly the latter, too detailed, sophisticated and intrusive going far beyond what was needed for the verification of a test ban treaty.[63] In the event, there was no final agreement on the details of the treaty's on-site inspection provisions.

The Separate Verification Agreement (SVA)

Despite the disagreements over OSI, the USSR accepted in principle the US proposal for a Separate Verification Agreement (SVA) in February 1978. The SVA was designed to meet a critical UK/US negotiating objective; namely, effective verification of the Soviet Union. Exactly what this meant was never really made clear, but at a minimum, this evidently needed to contain arrangements that would improve the detection and identification capabilities of seismic verification methods that made the prospect of Soviet cheating that would provide a military advantage going undetected appreciably less likely. Such a provision would be limited to the UK, US and USSR and would provide them with a more intensive regime that would apply to other states parties to the test ban treaty. The SVA's most novel feature would see the installation of National Seismic Stations (NSS) in the three territories to provide the other participants with instant information about seismic events. The role of such stations, however, drew upon concepts for an international monitoring network of stations placed around the globe that had first been proposed by the 1958

Geneva Conference of Experts.[64] Since this was an important requirement for the UK and the US, Soviet readiness to accept NSS appeared at the time to be a major breakthrough in the negotiations. The SVA also contained guaranteed and detailed rights and obligations for the conduct of an on-site inspection; that is, such provisions would not be at the discretion of the inspected state party, but a central part of the SVA itself and agreed before any inspection would take place. Finally, the SVA would also see a tripartite Joint Consultative Commission to monitor the practical working of the test ban treaty. Thus, at the end of the fourth round of negotiations on 21 March 1978, it seemed that the two sides might just be moving towards a treaty with an initial duration of four or five years with a review conference in the final year, which would have to grapple with various contentious issues – PNEs, duration and various verification issues.[65] But this was all based on the premise that an agreement could be reached on verification.

Despite all of this progress, and as will be discussed in detail in Chapter 4, the US and UK had second thoughts about the wisdom of a treaty of unlimited duration given a growing concern that this would entail too much risk to the continuing reliability of stockpiled nuclear warheads. A threshold treaty, which some in the US favoured to tackle this problem, was not attractive as that would run counter to UK objectives for a comprehensive test ban treaty and would virtually nullify its non-proliferation benefits and undermine Western arguments against the Soviet case in favour of continued PNEs in a test ban regime.[66] First the UK and then the US decided in April and May 1978, in favour of a treaty of five years' duration. Rather than complain and make political propaganda, the USSR were happy to go along with this since a limited-duration treaty likely suited their preoccupations with Chinese nuclear testing.

The idea of a testing moratorium fades from the scene and does not reappear as a serious prospect in the proceedings apart from a final flurry in July 1979 when there was a momentary anxiety that President Carter might feel obliged to call a moratorium on signature of a treaty, and this would impact on British testing plans in 1980.[67] MOD officials were certainly leery and wanted a thorough internal UK discussion before offering a public view on the matter.[68] From a defence nuclear perspective, there was always a strong case for being able to conduct nuclear tests until entry into force of a treaty.[69] Previously the Foreign Secretary had thought that it would be politically inconceivable to sign a treaty in a blaze of goodwill and then to go on testing.[70] British security interests, however, still lay in using all possible opportunities test in order to enhance AWRE's nuclear expertise.[71] This point never seems to have changed throughout the negotiations.

Notwithstanding this ostensible retreat from the original aspirations for an indefinite duration test ban treaty, there was sufficient common ground to enable the three delegations in Geneva to start detailed drafting of treaty text. By the summer of 1978, they produced an agreement on all but one of the treaty's main Articles and on the PNE protocol that would be an integral part of the treaty. David Owen felt that, without a Presidential push, and on the basis of current trends, the negotiations were unlikely to be completed before the end of September 1978.[72] The sole exception to the sense of progress was the mandate for the envisaged review conference that would take place in the final year of the Treaty's initial duration, and hence on the prospects for its future duration, for example, a further period of five years, indefinite extension, or simply its expiration. Intensive work continued in Geneva on all the verification provisions in the SVA. The US proposed 15 NSS at specific locations in the Soviet Union, but the issue of NSS was not so easily solved and ended up putting the UK on the spot of which more later in this chapter. National Seismic Stations were to become one of the main stumbling blocks to progress in the tripartite negotiations.

National Seismic Stations (NSS)

As noted earlier, the installation of national seismic stations on the territory of each of the other two state parties was a fundamental feature of the Separate Verification Arrangement (SVA).[73] These stations would enable the continuous exchange of authenticated seismic data, which was essential for the monitoring of the ban since it would help with the identification and discrimination of seismic events such as earthquakes and non-nuclear explosions, for example, mining and quarrying. These seismometers would be located in 100-metre bore holes and would be owned and manned by the host country, but they would be installed by representatives of the other SVA partners and would include authentication devices and tamper indicating seals. UK Ministers decided in April 1978 that the UK would only associate itself with the proposed ten US seismic stations to be situated in the Soviet Union and not seek ten stations of its own.[74] The US wanted to install US-made equipment on Soviet territory, but the Russians were not convinced of the need for this, no doubt harbouring suspicions about the capabilities and true purpose of American-made equipment. However, there were other unresolved problems with the NSS concept. First, the USSR said that it would accept ten stations as proposed by the US (and supported by the UK), provided that there were also ten in the US, which was agreed at once, and ten in the UK and its dependent territories. However, the UK agreed to one – at its existing seismic station at Eskdalemuir in Scotland, which had become fully operational in 1962 as a central part of the UK's forensic seismic research programme for nuclear test monitoring. In response, the Russians insisted on ten, which was, in the British view, an absolute technical absurdity. Sadly this now enabled the Russians to stall on the negotiations for over 18 months. By mid-1980 that stalemated condition would remain until the US Presidential election in November that year.[75] The Soviet invasion of Afghanistan on 26 December 1979 and President Carter's subsequent request to the Senate on 2 January 1980 to postpone its consideration of the Strategic Arms Limitation Treaty II (SALT) provide a reminder of the wider poor and declining state of East-West relations in which the test ban treaty was being negotiated. US-Soviet relations were already at a low ebb, and Afghanistan just compounded matters. On the direct impact on the test ban negotiations, the US decision was to proceed with them, but 'at a slow pace'. A test ban treaty could not come before SALT II, meaning that the Carter Administration would not look for an early completion of a test ban treaty.[76] That said, a comprehensive nuclear test ban treaty continued to be one of President Carter's most important objectives in the arms control field in 1980 as the Presidential Election approached. Carter still wanted to seek to impose further qualitative constraints on the nuclear arms competition between the USSR and the US.[77] Elsewhere NATO's 'dual-track' decision in December 1979 to modernise its long-range theatre nuclear forces with the deployment of 464 new nuclear-armed cruise missiles and 108 Pershing II ballistic missiles in response to extensive Soviet deployments of its new SS-20 intermediate-range nuclear missile whilst simultaneously pursuing an arms control solution had generally relegated the importance of CTBT, and high-level political interest in it, as an arms control measure at that particular time.[78] The CTBT as the Holy Grail of arms control was becoming even more elusive.

Political equivalence between all parties in a treaty ostensibly mattered a very great deal to the Russians. They admitted that there was indeed no technical basis for 10 NSS in the UK and its dependent territories in the same way that there was no technical basis for 10 in the USSR. Their case was purely a political one, namely that there should be equal obligations.[79] One could argue, however, that a much more likely explanation was that this offered a convenient excuse to avoid intrusive verification measures on Soviet territory. Second, the installation timetable for the NSS raised obstacles too. Washington wanted all the NSS to be in place and operating within the first two years of the treaty. London supported this, but UK officials had doubts on the value and feasibility of this objective in a treaty limited to three years. It certainly would be a tall order and a very great expense to install and commission ten stations in two years. Third, the US had very high performance standards for the NSS, which were not unreasonable if the treaty's verification regime were to be effective and for it to withstand Congressional scrutiny. Washington required that the seismometers and their ancillary electronic equipment had to meet very high US technical specifications; and, furthermore, the Americans wanted the data from the stations to be authenticated, and continuously transmitted without delay, preferably via satellite. Although the Russians agreed to some discussion of system performance, they refused to discuss the other important technical aspects until their proposal for 10 NSS on UK territory was agreed. In addition, the Russians refused to entertain the use of satellites, arguing instead for much slower means of communication (perhaps transfer of recording tapes by ground and air), which would delay the transfer of data by several weeks and increase the risk that the tapes might be damaged, or lost in transit. That would certainly diminish the utility of what was intended to be a real-time detection and reporting system. Finally, as per the UK and US NSS proposal, the stations would be installed and maintained by UK/US teams in order to ensure their reliable operation. The USSR did not accept this proposal either.

In what turned out to be the final year of the negotiations, the NSS issue assumed a disproportionate role and consumed a good deal, if indeed not all, of UK official and Ministerial time devoted to test ban issues, especially since the US started to press the UK repeatedly in 1980 to see whether it could show some flexibility on the number of NSSs it would be willing to host. John Edmonds had noted, in fact, that the NSS/UK issue had distorted Ministerial consideration of CTBT issues for nearly two years.[80] Washington was disappointed, to say the least, that the UK could not offer more flexibility and accept a few more stations in order to secure an agreement on a treaty.[81] We now need to look in greater detail at just how the UK adopted and maintained the position it did – in short it was one of both principle and science. And both were totally independent of any need to find an excuse to avoid a test ban treaty and its potential adverse impact on the UK's own nuclear weapons programme as a decision approached in 1980 on whether to acquire a new strategic weapon system in succession to Polaris/Chevaline. Indeed the US delegation in Geneva did wonder whether a delay in the UK's decision on NSS was somehow linked to a more fundamental doubt about the wisdom of a CTBT. They were also aware of the new Prime Minister Margaret Thatcher's sceptical attitude to a treaty as reported to them following her meeting with the US Secretary of State in June 1979.[82]

First consideration of the ten NSS proposal

The UK had its first detailed consideration of the Soviet insistence that there should be ten NSSs on UK territory in January 1979.[83] The Cabinet Office-chaired interdepartmental Official Group on International Aspects of Nuclear Defence Policy (GEN 63) met to consider a draft Foreign Office submission to Ministers on what the UK reaction to the Soviet proposal should be.[84] Officials were keen to avoid reconvening at the next session of the tripartite negotiations on 29 January without being in a position to make a formal response to the Russians. The FCO's draft paper considered several options, but the Official Group felt that the most constructive response, and probably the least the Americans would expect the UK to make, might be to say in principle that the UK had no objection to NSS on UK territory, or on the dependent territories subject to satisfactory agreements on costs, locations and certain other technical aspects. Britain could say that it had examined all possible locations against the criteria of political or constitutional acceptability, technical value and logistic practicability, but only three sites satisfied all criteria – Eskdalemuir in Scotland and two other sites in the Caribbean. However, the UK would have to be careful to avoid making any commitments before a Ministerial decision on costs. Ministers had already agreed in correspondence that the UK could accept at least one station in the UK and probably one or two in dependent territories. Moreover, Prime Minister James Callaghan felt strongly that the UK needed to be clear on departmental responsibility for costs before any final decisions were made. Whitehall departments were no doubt keen too to avoid having the costs coming off their own budgets unless new money was forthcoming. Neither the FCO nor MOD was able to fund those potentially very significant costs from within their current budgetary allocations.[85] And it would be Treasury practice to insist that any additional costs would have to be met from within existing budgets. Furthermore, the UK was also keen to avoid going beyond three stations as that could not be justified technically in terms of contributing to an effective verification regime. Ministers and their departments would certainly not want to have to spend money on technically unjustified stations that met no treaty verification need, especially given the dire domestic economic conditions at that time and resultant pressures on the public purse.[86] In any case, there was a better technical case for offering an array station rather than a single borehole as per the NSS concept on an island site, which would enable the UK to provide fewer but more capable sites.[87] Eskdalemuir was an array station, and it was this British concept that had proved very effective in teleseismic monitoring of Soviet testing at the Polygon test range at Semipalatinsk.[88] There was, furthermore, a better case for an NSS in Commonwealth countries such as Australia. Officials thought that both these options might be pursued, but there would be a risk that to do so would give the Russians more scope to undermine the general case for the NSS concept.

Perhaps inevitably the Cabinet Office-chaired Official Group concluded that more work was needed before it would be in a position to put a paper with clear recommendations to Ministers on the NSS question. Ministers would certainly have to meet to discuss the financial implications and might have to consider the fundamental question of whether UK membership of the SVA was worth the price that the UK would have to pay. Officials thought that the next step would be for the UK to discuss the problem further with the Americans at a planned bilateral meeting scheduled for Washington the following week on 17–18 January 1979. The UK would certainly be prepared to discuss the question further with the Russians in order to try to discover the minimum position they would accept. If this were more than three sites, then the financial cost to the UK could not be justified on technical grounds, but would need political justification and officials would need to consult Ministers again. Officials would want to remind the Americans of the mutual advantage of UK membership of the SVA and try to establish the extent to which they might be prepared to help meet UK costs. Furthermore, a key aim for the Washington meeting would be to establish what might be the minimum subscription the UK would have to pay for its membership of the SVA and the maximum help that it could expect from the Americans. In light of the outcome of that Washington meeting, the Group tasked FCO to prepare a revised submission to Ministers to enable them to decide whether the resultant cost to the UK was a price worth paying and how it should be funded. This was done and the Cabinet Official Group convened again on 22 January 1979 to review the FCO draft.[89] The key point now seems to have been that Ministers would not need to be asked to take a decision on accepting as many as 10 UK NSS, even in principle. In its response to the Russians, the UK would not need to specify a number, but should note that it had no objection in principle to UK NSS, it had established that there were only three sites that were constitutionally and technically acceptable. Once the tripartite negotiations resumed, the UK could then probe the Soviet position and report back to Ministers before a final view was taken on number and locations. There were no other really satisfactory options. Costs continued to be a major complicating factor, in part because these were still so uncertain, even US estimates were far from being definitive. At worst the UK could be looking at installation costs of £30 million for ten NSS with annual running costs of as much as £15 million. These eye-watering figures did not include subsequent data processing and handling costs, which might amount to a few more million each year.[90] And nor could US officials offer any guarantee to meet some or all of the UK costs as that would be in the sole gift of Congress. The FCO paper was amended in light of the Group's subsequent discussions and circulated for clearance by correspondence prior to submission to the Cabinet Secretary Sir John Hunt. The aim was to secure a Ministerial view before the tripartite negotiations resumed, which was likely to be on 7 February 1979. Ministers were invited to adopt one of two positions:

1) To argue that the UK's role in the SVA was only in association with the US; that the UK should not therefore be forced to accept 'equal obligations'; that 10 UK NSS were technically unjustified; and that the UK network should consist of no more than two or three stations. If this approach did not move the Russians, then the UK could withdraw its bid for membership of the SVA; or

2) That the UK should try to probe the Soviet position and enable further submission to be made to Ministers for a decision. However, like option (1), it would still involve the UK accepting in principle at least three UK NSSs.[91]

Ministers finally decided that the UK was only prepared to accept one NSS as the only appropriate number.[92] Publicly at least the Foreign Secretary was still optimistic, seeing no reason why the outstanding issues should not be resolved quickly when the negotiations resumed on 29 January.[93]

The negotiations ended up in stalemate ostensibly as a result of this British position. The Russians had no incentive to move from the 10 NSS as per their original demand. Unfortunately for the UK, US officials had started to express increasing impatience with the British position and had indicated in Geneva that the minimum the UK should offer was about six NSS.[94] FCO, MOD and Cabinet Office officials recognised that Ministers had taken a firm decision to offer only one NSS in the UK and, if they were to be asked to reconsider this, they would first need to be convinced that there was no prospect of persuading the Russians to accept the UK position. If the UK were to reconsider its position, it would need to take account of the degree of importance to be accorded to UK participation in the SVA and to its support for the US on NSS, the financial aspects and whether UK agreement to technically unjustified NSS would undermine the verification of future arms control agreements.

The Official Group was clearly vexed by US attitudes to the UK approach on NSS, especially since the US had been unable since December 1977 to clarify its own position on 'permitted experiments',[95] and had not yet produced treaty Review Conference language that looked like being remotely negotiable with the Russians. Both of these were as important as the NSS issue in blocking progress. The Americans intended once a test ban treaty had entered into force to conduct very small nuclear experiments (of yields below 100 lb TNT) in order to maintain their technical capability for stockpile maintenance. Such experiments are not nuclear tests in the accepted sense of the term and as such, at least in the UK view, would not detract from a test ban treaty. Indeed the UK had similar requirements, but no definite plans were yet in place. Experiments of this type could not be used to test weapons in the stockpile, or to develop new ones. The US wanted some understanding with the Russians that these sorts of experiments did not fall under the treaty's prohibitions.[96] This issue is discussed in greater detail in Chapter 4.

In the Soviet view, Review Conference language would determine the verification measures for the Treaty. Thus, in British eyes, the UK position on accepting only one NSS could not be presented as the sole obstacle to progress in the treaty negotiations; or that if the UK were to change its view, rapid progress would magically ensue on all the other contentious issues such as on-site inspection. As to claims by US officials that the UK was undermining their position on NSS, UK acceptance of 10 NSS would be tantamount to accepting that NSS were, as the Russians claimed, purely a political issue. For now, the UK would maintain its position until the recess in the negotiations, which was scheduled for 7 April 1979. The gist of this view was conveyed to the US delegation in Geneva; in short, the UK was not going to change its position and that if were to do so, this would not likely unblock the negotiations given the US problems over 'permitted experiments' and Review Conference language. UK officials had hoped that the US might be willing and able to provide funds for NSS on UK territory, but this proved impossible on political grounds in Washington;[97] Congress simply would not wear it. Nevertheless, American officials continued to urge the UK to move 'more than half way' towards the Soviet demand for ten NSS in the UK and dependent territories as the only way of breaking the then impasse in the negotiations. Washington's view was that the UK should offer (and pay for) six NSS.[98] That said, President Carter had written to Soviet President Brezhnev in June urging him to stop obstructing progress by the unrealistic demand for ten NSS.[99] But that did not make any dent in Soviet obduracy.

The new Conservative government decided in September 1979 that when the tripartite negotiations resumed on 24 September, the UK delegation would confirm to the Russians that the previous government's offer of one NSS in the UK itself at Eskdalemuir still stood.[100] London informed the Americans of this decision first before telling the Russians. Washington embassy officials were instructed to tell the State Department that if the UK failed to persuade the Russians to accept this offer, the most probable alternative was for the UK to withdraw from the SVA, but hoped that it would not come to that. Britain still had an interest in achieving an early and satisfactory test ban treaty, but it was also convinced that the ten NSS in the UK-dependent territories were not a decisive factor in the overall Soviet attitude to a treaty, nor that by accepting more NSS would such a concession lead to early progress on the other outstanding thorny issues in the negotiations such as permitted experiments, Review Conference language, and equipment to be installed in NSS.[101,102] The Russians were simply not impressed by this reasoning and continued to sit on their position. Inevitably, therefore, this particular problem remained unresolved for the rest of 1979. The Prime Minister told President Carter in response to his question on whether there was any further flexibility on the NSS that the UK could only afford one NSS and if this were an insuperable difficulty, then the UK was prepared to withdraw from the negotiations.[103] Cyrus Vance, US Secretary of State, then raised the question of NSS again with Lord Carrington on 18 December 1979. He asked whether the UK could accept three additional NSS, if the US could provide the funds, in order to break the deadlock in the negotiations. Carrington said that the UK's difficulty had been in justifying to Parliament expenditure on redundant NSS, but he thought that the UK could find places such as Ascension Island for three additional stations. Vance stressed, however, that he had no authority yet for US funding support, but that if he were to look afresh at the finances, he hoped the UK would look at locations.[104] Nothing came of this, it was just one of those ideas aired in a negotiation or side conversation that gains no traction and quietly withers away. As John Edmonds was to remark in May 1981 from the moment in November 1978 that the Russians proposed that there should be ten seismic stations in the UK and dependent territories, this was virtually the only issue in the negotiations that Labour and Conservative Ministers devoted any serious attention to – primarily because of the millions of pounds sterling that such stations could entail.[105] A statement that goes to show that economics is more than likely to compromise arms control requirements in Ministerial minds at critical moments. Even so, the UK could not arrive at a position that was either negotiable with the Russians, or which satisfied the Americans.

Stalemate

As the tripartite negotiations marked time in 1979 and into 1980, there seemed to be little or no prospect that any of the contentious issues could be resolved. Divisions in Washington were acute too between the doves and hawks, with the latter not at all keen on even a three-year treaty. Indeed, Herbert York, the new US ambassador to the negotiations, had told the Cabinet Official Group on the International Aspects of Nuclear Defence Policy in mid-February 1979 that there had long been a substantial division, which existed all the way up to and immediately below the President between those that favoured a test ban treaty and those who did not. York thought the opponents probably formed a majority.[106] John Edmonds offered some options for solutions to the Geneva stalemate in his despatch of 28 July 1981 to Lord Carrington that looked retrospectively on the negotiations. A crucial part of Edmond's assessment was that there had been an urgent need for the US to change aspects of its negotiating position if there were to be any chance of progress, and in particular on its attitude to the future of a test ban after its three-year duration.[107] Much would depend on the outcome of the US Presidential election in November 1980, which made it difficult for any UK policy review in the interim. As early as April 1980, it seemed that it was difficult to imagine that a CTBT would be completed in 1981, and if Ronald Reagan (the Republican candidate) were elected it would be more than likely that the treaty would be scrapped. Even if President Carter were to be re-elected, it was unlikely that he would press ahead as earnestly as he had done at the start of his Administration in 1977 given the deterioration in East-West relations and all the internal grief that the test ban had caused in the inter-agency process.[108] As ever in US politics, as a change in Administration loomed in an election year, policy-making in Washington would grind to a halt with little prospect of any new major or even minor initiative, or modified negotiating proposal emerging from the election year stasis. Something inevitable really, especially in a highly contested area such as a test ban treaty. Prospects for a treaty were now vanishing fast. A Cabinet Office official snidely remarked on Edmonds' despatch, which was dated 28 July 1980, a fortnight after the government's decision to acquire Trident C4 submarine-launched ballistic missiles from the US was announced, showed no awareness of the changed context. The Trident decision implied a compelling need for some UK tests to develop and validate a new nuclear warhead. Such developments were, as the official remarked, 'likely to be close to Mrs T's heart, but the despatch does not even acknowledge their existence'.[109] In other words, advice from Geneva on the desirability of a test ban and how the UK could help bring it about were likely to fall on very stony ground indeed.

Ronald Reagan was elected President of the US in November 1980. The tripartite negotiations were suspended in mid-November with no date for resumption set pending a US policy review on the desirability or otherwise of a test ban treaty. Given the policies of the Reagan Administration, the outcome of such a review would likely come out against a treaty, which certainly had been the expectation in the FCO earlier in the year as already noted. In January 1981 the Official Group on the International Aspects of Nuclear Defence Policy met to consider the prospects for the treaty negotiations, to look at the possibilities and implications of a low threshold test ban (TTB), and to consider the nature and content of a policy paper for Ministers to consider.[110] The FCO and MOD had started a review of UK policy back in the autumn of 1980 as part of its preparations for the outcome of the US Presidential elections, and a TTB had been one of the key themes in that review of options with the MOD indicating a preference for a threshold of 5 kilotons.[111] However, a TTB was not seen as a good alternative, it had zero appeal as a non-proliferation measure and created just as many, if not more, verification challenges than an outright ban – discrimination and yield estimation to name but two. There were no compelling reasons to switch to a TTB. Officials thought that the Strategic Arms Limitation Treaty (SALT) and theatre nuclear forces would be higher priorities for the US than a test ban treaty and the prime focus for the Reagan Administration would be on satisfying defence rather than arms control requirements. Moreover, a US policy review was unlikely to be settled for six months. As for a low threshold TTB, parts of the analysis appeared to suggest that UK defence requirements could be met under such a treaty with a threshold lower than 30 kilotons. However, this possibility was conditional on the achievement of improvements on testing technology, or on obtaining access to relevant US warhead secondary designs (i.e. the thermonuclear component of a two-stage nuclear weapon).[112] In the latter case, there was some doubt as to whether a primary of UK design could be matched to a secondary of US design. Technical factors apart, reliance on secondary mechanisms of US design – assuming that they would be made available to the UK – would increase the UK's dependence on the US. Although nuclear testing, associated with the UK's planned nuclear weapons development programmes could probably be fitted into a three-year period after 1986, this did not mean that a 30-kiloton threshold could be lowered at the end of that period. Unforeseen developments in conventional defences might give rise to further requirements for new weapons development and thus to the need to conduct nuclear tests again and it was not safe to assume that these could be carried out with a threshold of less than 30 kilotons. The prospects for moving in time to a lower threshold depended on progress in the development of testing technology. Presumably, this was an allusion to enhanced test diagnostics, which would provide much more data on the performance of the test device. Officials agreed that further work on a policy submission for Ministers would be required and this work would be a joint effort by the FCO and MOD and would take the form of a memorandum for the Foreign Secretary to circulate to key colleagues. Essentially the choice for Ministers in terms of determining British policy on a test ban fell between making a choice between two options. It would be either a case of maintaining the freedom to conduct nuclear tests, which was required if the UK's nuclear weapons programme was to be preserved, either by abandoning the test ban negotiations altogether in company with the US, or just possibly by seeking to negotiate a threshold test ban treaty instead. Alternatively, the UK could continue to negotiate for a three-year test ban treaty with the risk that it could prove politically very damaging to resume nuclear testing at the end of the three-year period.[113] Officials recognised very clearly that it would ultimately be the position of the US that would determine the future of the negotiations. As it transpired, this proved to be a prescient if rather obvious observation on the political realities facing the UK as the junior partner in these proceedings.

By May 1981 the UK was feeling the heat and felt that it could not ignore the pressure in the multilateral Committee on Disarmament in Geneva on nuclear testing where many member states saw progress on a treaty as a litmus test of the nuclear weapon states commitment to arms control.[114] The US and UK were blocking CD efforts to establish its own working group on CTBT issues, which evidently did not help the atmospherics. Despite this, Ministers were at least still publicly committed to a test ban as a worthwhile objective. FCO officials hoped to convince the US to advance their decision-making on this issue and certainly not to take any decisions without prior consultations with the UK. London saw three options at this time, and all of them had problems; abandoning the negotiations would entail significant political costs and damage arms control and non-proliferation interests, sticking with the then-current three-year treaty would almost certainly result in pressure to extend it on its expiry, and going for a threshold treaty would have no non-proliferation advantages. We should note here that the principal reason for the failure of the Second NPT Review Conference in 1980 to produce a consensus Final Declaration was the bitter dissatisfaction that many state parties had with the failure of the nuclear weapon states to agree a CTBT.[115] In addition, to trying to convince Washington to focus on the test ban, London wanted to put UK proposals on how things might be handled to the US before it settled its position.[116] The major concern for the UK here, as ever, would be that once the Washington inter-agency had finally agreed a position and then gone public with it, it would be extremely difficult, if not impossible to change it again. This was especially true where the topic was as highly contentious and divisive in the interagency process as the test ban. By January 1982 the US had not made as much progress with its policy review as it had hoped, but US officials had told their British counterparts that they now regarded a test ban treaty as, at best, a long-term objective for arms control.[117] Like the UK, the US believed that in the near to medium term a treaty would damage Western security interests, a point not lost on the Prime Minister. Any test ban treaty, if it were to be compatible with current British plans, had to allow for a development of a warhead for Trident and for a replacement for the WE177 after that. This was still the UK core position in early 1982, which shows that the UK was still some way from having a tested warhead design for Trident ready to enter production. This is discussed further in Chapter 5. For one senior Cabinet Office official dealing with nuclear policy at least – Dr Robert Press – a threshold treaty probably offered the least damaging option open to the UK in the dilemma it faced between its declared political objectives for a test ban and giving due concern to its defence nuclear interests.[118] Unfortunately, the US had by this stage probably widened the gap between the threshold level that they and the UK might be prepared to settle for – a development that likely scuppered the prospects for a threshold test ban treaty as any sort of viable negotiating option with the Soviet Union.

On 21 January 1982, Lord Carrington finally proposed to the Prime Minister a new approach to the negotiation of a test ban treaty that drew upon an earlier agreed set of options between FCO and MOD officials.[119] Carrington's new approach contained four options: inaction, leaving it to the US to explain the inaction in Geneva; a joint announcement of suspension explaining that a test ban treaty was a long-term objective with the focus for now on START and INF; active presentation that a test ban treaty was incompatible with UK defence interests; or a step-by-step approach to testing restrictions aimed at retaining some momentum towards a treaty whilst safeguarding UK testing plans. The step-by-step approach to a treaty with a comprehensive ban being a long-term rather than an immediate objective was the preferred option offered by officials. Perhaps surprisingly the Prime Minister agreed that this idea should be put to the Americans.[120] This new approach was shared with the Americans during a visit from David Gillmore, Assistant Under-Secretary (Defence) in the FCO to Washington at the end of January, and US reactions were less negative than he had feared.[121] That said, the JCS were still isolated in their opposition to any work in the CD on testing issues.[122] The issue then went to the White House for a decision.[123] President Reagan approved an approach broadly similar to the one suggested by the UK during discussions between US and UK officials in January and February, namely placing the immediate focus on verification and compliance issues.[124] A test ban treaty was now a long-term objective, a view that did not go down well with the US academic and non-governmental arms control community.[125] The US announced this position to the CD on 9 February, noting that while a comprehensive ban on nuclear testing remained an objective, the ban would not be of assistance in present circumstances in reducing the threat of nuclear weapons, or in maintaining the stability of the nuclear balance.[126] Washington also agreed that, in tactical and procedural terms, the best thing to do now would be to seek agreement in the Committee on Disarmament (CD) in Geneva to the establishment of a new sub-group whose mandate would be carefully framed to ensure that its focus did indeed remain on these particular aspects. Pressure from the UK led Washington to agree setting up such a working group, a move which was driven out of interest in maintaining some continuity in UK policy and out of concern for the impact on public opinion at home and abroad if the Americans were to abandon a CTBT completely.[127] The UK and the US had nevertheless previously rejected a Group of 21 neutral and non-aligned member states proposal to establish a working group in the CD with a negotiating mandate to produce a test ban treaty.[128] Despite this, and somewhat surprisingly, the Group of 21 did agree to the joint UK-US proposals; and then eventually the Soviet Union and its Eastern Group allies also bit the bullet on the last day of the CD's Spring 1982 session and supported it too.[129] A working group was thus established at the end of April. By this time British Ministerial attention was, in any case, fully engaged on the Falklands War.[130] The UK had played a key part in setting this working group up and then went to play an active part in its work and with the associated Group of Scientific Experts that was looking in detail on the seismic verification requirements for a future treaty.[131] And that focus on seismic verification requirements would be the essential focus for the CD until 1994 when a negotiating mandate for a test ban treaty was agreed following an initiative of a new US Democratic President. In the meantime, the UK view remained that the ability of the sort of multilateral treaty verification regime under consideration in Geneva fell short of that required to detect military significant cheating notwithstanding the views of academic seismologists who argued that a treaty could be effectively monitored.[132]

Conclusion

In retrospect, the tripartite negotiations never had much chance of success in producing a truly comprehensive nuclear test ban treaty of indefinite duration. One of the principal problems was the US's and UK's fears over the impact of a treaty on nuclear weapons stockpile safety and reliability allied to internecine squabbling in Washington on the merits and demerits of a treaty. Once a short-duration treaty became the preference, then the prospects for a truly comprehensive test ban treaty vanished. A threshold treaty would be irrelevant from a non-proliferation perspective and raised significant verification challenges of its own. This issue also made it even more difficult to forecast how the negotiations might develop and would likely change their whole shape.[133] In addition, Soviet attitudes to verification, especially their technically absurd demand for 10 NSS on UK territory and dependent territories, only raised the barrier to a treaty higher still and effectively gummed up the negotiations. The UK need to develop and test a warhead for Trident after the July 1980 decision certainly moderated UK enthusiasm for a treaty of indefinite duration. Furthermore, with Margaret Thatcher as Prime Minister, who had on her election already expressed her scepticism about the desirability of a test ban treaty, the chances receded further off into the future. And as we have seen, the Reagan Administration had decided that a test ban treaty was not in US security interests and only a long-term objective – an outcome that suited the UK perfectly. Mrs Thatcher's government had no trouble with this position.[134]

Notes

1 Thomas A. Halsted, 'Why No End to Nuclear Testing?', *Survival*, Vol. 19, No. 2, 1977.
2 See British NPT state papers at RUSI Marking the 50th Anniversary of the Nuclear Non-Proliferation Treaty, Royal United Services Institute (rusi.org).
3 *United States Nuclear Tests July 1945 Through September 1992*, U.S. Department of Energy, National Nuclear Security Administration, Nevada Field Office, DOE/NV-209-REV 16, September 2015, https://www.nnss.gov/docs/docs_LibraryPublications/DOE_NV-209_Rev16.pdf. page xiii.
4 V. N. Mikhailov, editor, *USSR Nuclear Weapons Tests and Peaceful Nuclear Explosions: 1949 Through 1990*, Sarov, Russia: The Ministry of the Russian Federation for Atomic Energy, and Ministry of Defense of the Russian Federation, 1996.
5 TNA FCO 66/898, History of Nuclear Test Ban Negotiations: 1958–1976, ACDRU, 26 May 1977.
6 TNA FCO 66/616, D.M. Summerhayes, Arms Control and Disarmament Department to Mr Thomson, Mr Coles, Private Secretary, Threshold Agreement on the Conduct of Underground Nuclear Tests, 10 June 1974.
7 TNA FCO 66/615, D.M. Summerhayes, Arms Control and Disarmament Department to Mr Pellew, Defence Department, Mr Montgomery, Threshold Test Ban – Possible US/Soviet Initiative, 8 May 1974.
8 TNA FCO 66/615, M.E. Pellew, Defence Department to Mr Montgomery, ACDD, 22 May 1974.
9 TNA FCO 66/616, A.F. Hockaday, DUS (Policy and Programmes), MOD to J.A. Thomson, FCO, Threshold Agreement on the Conduct of Underground Nuclear Tests, 6 June 1974.

10 TNA FCO 66/617, B. Richards, Arms Control and Disarmament Department to Mr Thomson and Parliamentary Unit, Banning of Nuclear Tests – PQ to the Prime Minister by Mr William Molloy: 11 July 1974, 9 July 1974.

11 TNA FCO 66/617, B. Richards, Arms Control and Disarmament Department to Mr Thomson and Parliamentary Unit. Banning of Nuclear Tests – PQ to the Prime Minister by Mr William Molloy: 11 July 1974, Background, 9 July 1974.

12 See Chapter Six in John R Walker, *British Nuclear Weapons and Test Ban 1954–1973: Britain, the United States, Weapons Policies and Nuclear Testing: Tensions and Contradictions*, Farnham, Ashgate, 2010.

13 See also United Kingdom Working Paper on the Estimation of Depth of Seismic Events, CCD/402, 1973.

14 TNA FCO 66/618, Agenda Item 30, Urgent Need for Suspension of Nuclear and Thermonuclear Tests, 1974.

15 TNA FCO 66/774, Test Ban Issues, Agenda Item 122, 30th Session of the UN General Assembly, Urgent Need for the Cessation of Nuclear and Thermonuclear Tests and Conclusion of a Treaty Designed to Achieve a Comprehensive Test Ban, Arms Control and Disarmament Department, FCO, IOC (75) 212, 30 October 1975.

16 United Nations General Assembly, Thirty-first Session First Committee, Agenda item 47, Conclusion of a Treaty on the Complete and General Prohibition of Nuclear Weapon Tests, A/C.1/31/9, 22 November 1976.

17 TNA DEFE 24/1344, Record of a Meeting on a Comprehensive Test Ban (CTB) Held in Washington on 14 March 1977; TNA DEFE 24/1344, Record of a Meeting with the Director of ACDA, Monday 14 March 1977.

18 Thomas A. Halsted, 'Why No End to Nuclear Testing?', *Survival*, Vol. 19, No. 2, 1977.

19 TNA DEFE 19/242, Percy Craddock, Head of UK Delegation to the CTB Negotiations, FCO to David Owen, Negotiations for a Comprehensive Ban on Nuclear Tests: Part I (Narrative), 21 April 1978.

20 TNA DEFE 24/1344, Frank Cooper to Secretary of State, Nuclear Arms Control, 4 May 1977.

21 Hansard, HC Deb, Vol.933 col.558, 16 June 1977.

22 TNA FCO 46/1566, Draft Brief, Visit of the Prime Minister to Washington 9–11 March 1977, Nuclear Testing, Brief by the Foreign and Commonwealth Office, PMVE (77), 24 February 1977.

23 TNA DEFE 24/1344, J.C. Edmonds, Arms Control and Disarmament Department to Sir R. Sykes, Nuclear Testing, 22 March 1977.

24 TNA DEFE 24/1344, H. Bondi, Chief Scientific Advisor to Secretary of State, Nuclear Arms Control, 6 May 1977.

25 TNA FCO 66/897, E.A.J. Fergusson to Patrick Wright, 10 Downing Street, Nuclear Arms Control, 5 May 1977; TNA DEFE 24/1344, Brief for the use by the Prime Minister with President Carter, Test Ban Negotiations, 9 May 1977.

26 TNA DEFE 24/1344, R.L.L. Facer, Private Secretary to PS/CSA, Nuclear Arms Control, 9 May 1977.

27 TNA DEFE 24/1344, Clive Rose to Mr Wright, Test Ban Negotiations, 6 May 1977.

28 TNA DEFE 24/1344, W.K. Prendergast, FCO to Patrick Wright, 10 Downing Street, Comprehensive Test Ban, and enclosed draft letter to President Carter, 11 May 1977.

29 TNA FCO 66/898, FCO Telno. 1439 to Washington, Mr Brezhnev's Message to the Prime Minister, 31 May 1977; TNA FCO 66/898, FCO Telno. 1436, Text of President Carter's Message to Prime Minister, 31 May 1977; TNA FCO 66/898, GEN 63 (77) 4, Cabinet International Aspects of Nuclear Defence Policy, Test Ban Negotiations, Note by the Chairman, 18 May 1977.

30 There are a very large number of FCO Defence Department files on SALT covering the years 1969–1979 available in The National Archives that show the extent of British lobbying and anxiety on non-transfer and non-circumvention – see for example one year's worth of files in FCO 46/1001 to FCO 46/1006 Strategic Arms Limitation

Talks between the Soviet Union and USA 1973; see also John R. Walker, *Britain and Disarmament: The UK and Nuclear, Biological and Chemical Weapons and Arms Control and Programmes 1956–1975*, Farnham, Ashgate, 2012.

31 TNA FCO 66/876, Anglo/United States Consultations on a Comprehensive Test Ban Treaty (CTBT), Washington, 5–6 July 1977, Steering Brief, Foreign and Commonwealth Office, 30 June 1977.

32 The negotiations were conducted very formally and rigidly with only the heads of delegations speaking at plenary meetings. Delegation heads, moreover, usually met privately beforehand to agree the topics that would be discussed. The serried ranks of experts sat mute in the back rows.

33 John R. Walker, *British Nuclear Weapons and the Test Ban 1954–1973: Britain, the United States, Weapons Policies and Nuclear Testing: Tensions and Contradictions*, Farnham, Ashgate, 2010.

34 TNA FCO 46/1567, W.J.A. Wilberforce, Defence Department to Private Secretary, 22 July 1977.

35 TNA PREM 19/41, Cabinet Secretary's Incoming Brief for new PM, Comprehensive Test Ban, May 1979.

36 TNA FCO 66/876, Anglo/United States Consultations on a Comprehensive Test Ban Treaty (CTBT), Washington, 5–6 July 1977, Steering Brief, Foreign and Commonwealth Office, 30 June 1977.

37 TNA FCO 66/876, Anglo/United States Consultations on a Comprehensive Test Ban Treaty (CTBT), Washington, 5–6 July 1977, Steering Brief, Foreign and Commonwealth Office, 30 June 1977.

38 Hansard, HC Deb, Vol. 940 cc258–6029, November 1977.

39 TNA DEFE 13/2469, B.G. Cartledge, 10 Downing Street to E.A.J. Fergusson, Foreign and Commonwealth Office, Comprehensive Test Ban Treaty, 11 July 1977.

40 TNA 66/875, (draft) Tripartite Consultations on Comprehensive Test Ban Treaty (CTBT), Geneva, 13 July 1977, Brief Number 9: Moratorium, Foreign and Commonwealth Office, 30 June 1977.

41 TNA FCO 66/875, Draft Anglo/United States Consultations on a Comprehensive Test Ban Treaty (CTBT), Washington, July 1977, 22 June 1977.

42 TNA FCO 66/899, Anglo/US consultations on Comprehensive Test Ban Treaty (CTBT): Washington, 3 June 1977.

43 TNA DEFE 13/2469, B.G. Cartledge, 10 Downing Street to E.A.J. Fergusson, Foreign and Commonwealth Office, Comprehensive Test Ban Treaty, 11 July 1977.

44 TNA FCO 66/875, Annex A, Anglo/United States Consultations on a Comprehensive Test Ban Treaty (CTBT), Washington, July 1977, 22 June 1977, Brief No.3: Verification: On-site Inspection. See also John R. Walker, *The CTBT: Verification and Deterrence*, VERTIC Brief 16, October 2011.

45 TNA FCO 66/900, B.G. Cartledge, 10 Downing Street to E.A.J. Fergusson, Foreign and Commonwealth Office, Comprehensive Test Ban Treaty, 11 July 1977.

46 TNA FCO 66/875, Draft Anglo/United States Consultations on a Comprehensive Test Ban Treaty (CTBT), Washington, July 1977, 22 June 1977; Brief No.3. Verification: On-site Inspection, Foreign and Commonwealth Office, 24 June 1977. See also John R. Walker, *British Nuclear Weapons and the Test Ban 1954–1973: Britain, the United States, Weapons Policies and Nuclear Testing: Tensions and Contradictions*, Farnham, Ashgate, 2010, pp 288–289.

47 TNA DEFE 24/1344, D.C. Fakley, Director D Sc 6 to M.G. MacDonald, ACDD, FCO, Comprehensive Test Ban: Verification, 25 April 1977. The UK continued to report on its seismic research programme at Blacknest with its findings in national Working Papers submitted to the Conference of the Committee on Disarmament: on 12 April 1976 on the UK's contribution to research on seismological problems relating to underground nuclear tests (CCD/486); on 12 April 1976, on the processing and communication of seismic data to provide for national means of verifying a test ban (CCD/487); and also on

12 April 1976 on the recording and processing of P waves to provide seismograms suitable for discriminating between earthquakes and underground explosions (CCD/488).

48 TNA DEFE 24/1344, Record of a Conversation Held at the Old Executive Office Building, Washington, Comprehensive Test Ban (CTB), 16 March 1977.

49 TNA FCO 66/876, Tripartite Consultations on a Comprehensive Test Ban Treaty (CTBT): Geneva, 13 July 1977, Draft Brief No. 10: Advantages and Disadvantages of a Comprehensive Test Ban Treaty, Foreign and Commonwealth Office, 30 June 1977.

50 John Edmonds once told the author that he thought the UK delegation was much more closely united and integrated than the American one – a point noted by his predecessor too. (See TNA DEFE 19/242, Percy Craddock, Head of UK Delegation to the CTB Negotiations, FCO to David Owen, Negotiations for a Comprehensive Ban on Nuclear Tests: Part II (Comment), 21 April 1978.) Christopher Mallaby, Head of the FCO's Arms Control and Disarmament Department at the time made a similar point in his memoir – Christopher Mallaby, *Living the Cold War Memoirs of a British Diplomat*, Stroud, Amberley Publishing, 2017 – electronic version 2021. All telegrams to the FCO reporting events, or seeking instructions went out under the Ambassador's name as was, and is, Foreign Office practice, unlike the US were all the competing agencies represented in the delegation had their own separate reporting channels back to Washington. Edmonds noted that London's bureaucracy was better coordinated than Washington's. Almost invariably, the UK delegation's recommendations were dealt with urgently and efficiently so that the delegation had its instructions when needed. All too often, the Americans did not. TNA CAB 164/1564, Nuclear Test Ban Negotiations 1977–1980 Part I – Where Are We? J.C. Edmonds UK Delegation to the Nuclear Test Ban Negotiations, Geneva to The Right Honourable Lord Carrington, London, 28 July 1980. Percy Craddock, head of the UK delegation in the first year of the negotiations, had also commented that the US delegation was 'not notably united' and 'that there was more than the usual disarray amongst US agencies on the subject (of a test ban)', TNA DEFE 19/242, Percy Craddock, Head of UK Delegation to the CTB Negotiations, FCO to David Owen, Negotiations for a Comprehensive Ban on Nuclear Tests: Part I (Narrative), 21 April 1978.

51 TNA DEFE 19/242, Percy Craddock, Head of UK Delegation to the CTB Negotiations, FCO to David Owen, Negotiations for a Comprehensive Ban on Nuclear Tests: Part I (Narrative), 21 April 1978.

52 TNA CAB 164/1564, Nuclear Test Ban Negotiations 1977–1980 Part I – Where Are We? J.C. Edmonds UK Delegation to the Nuclear Test Ban Negotiations, Geneva to the Right Honourable Lord Carrington, London, 28 July 1980. See also TNA FCO 46/2286, A. Reeve, Arms Control and Disarmament Department to Mr P. H. Moberly, PS/Mr Hurd and PS, Comprehensive Test Ban (CTB), 12 August 1980.

53 TNA DEFE 19/242, Percy Craddock, Head of UK Delegation to the CTB Negotiations, FCO to David Owen, Negotiations for a Comprehensive Ban on Nuclear Tests: Part I (Narrative), 21 April 1978.

54 TNA DEFE 13/2469, John Hunt, Cabinet Office to B.G. Cartledge, 10 Downing Street, Military Nuclear Issues, 25 October 1977.

55 TNA FCO 66/901, UKMis Geneva Telno 571 to FCO, Comprehensive Test Ban: New Soviet Position, 3 November 1977; TNA DEFE 19/242, Percy Craddock, Head of UK Delegation to the CTB Negotiations, FCO to David Owen, Negotiations for a Comprehensive Ban on Nuclear Tests: Part I (Narrative), 21 April 1978.

56 TNA DEFE 19/242, Percy Craddock, Head of UK Delegation to the CTB Negotiations, FCO to David Owen, Negotiations for a Comprehensive Ban on Nuclear Tests: Part I (Narrative), 21 April 1978.

57 TNA FCO 66/1080, Bryan Cartledge, 10 Downing Street to W.K. Prendergast, FCO, Nuclear Matters, 2 February 1978; Bryan Cartledge, 10 Downing Street to W.K. Prendergast, FCO, Comprehensive Test Ban, 6 February 1978.

58 TNA DEFE 19/242, Percy Craddock, Head of UK Delegation to the CTB Negotiations, FCO to David Owen, Negotiations for a Comprehensive Ban on Nuclear Tests:

Part II (Comment), 21 April 1978. See also TNA FCO 66/888, Comprehensive Test Ban (CTB): Verification; On-site Inspections, 1977; TNA FCO 66/901, C.L.G. Mallaby to Sir A. Duff and Private Secretary, Comprehensive Test Ban (CTB) Negotiations: State of Play, 9 December 1977. Ministers had agreed on 1 December 1977 that the UK delegation need no longer insist on mandatory on-site inspections on the understanding that the Russians would agree to satisfactory procedures for voluntary on-site inspections.

59 Hansard HC Deb, Vol 941, c721W, 10 January 1978.

60 The author served as a Joint Task Leader for the CTBTO's On-site Inspection Operation Manual where work on drafting the manual started in 1997 and was still incomplete in 2022. Things move slowly in international negotiations.

61 TNA CAB 164/1564, Enclosure 1 to CTB Despatch of 28 July 1980, CTB Negotiations – State of Play on Main Issues, July 1980.

62 See, for example, *CTBTO On-site Inspection Workshop-23, The Further Development of the OSI Equipment Lists*, 7–11 November 2016; and CTBT Science and Technology Conference 2021, Development of the first comprehensive draft list of equipment for use during OSIs, Gregor Malich, Preparatory Commission, 1 July 2021. https://conferences.ctbto. org/event/7/contributions/1369/.

63 TNA DEFE 19/242, Percy Craddock, Head of UK Delegation to the CTB Negotiations, FCO to David Owen, Negotiations for a Comprehensive Ban on Nuclear Tests: Part I (Narrative), 21 April 1978.

64 Cmnd 551, Report of the Conference of Experts to Study the Methods of Detecting Violations of a possible Agreement on Suspension of Nuclear Tests, Geneva 1 July to 21 August, HMSO, London, 1958.

65 TNA DEFE 19/242, Percy Craddock, Head of UK Delegation to the CTB Negotiations, FCO to David Owen, Negotiations for a Comprehensive Ban on Nuclear Tests: Part I (Narrative), 21 April 1978.

66 TNA DEFE 19/242, Percy Craddock, Head of UK Delegation to the CTB Negotiations, FCO to David Owen, Negotiations for a Comprehensive Ban on Nuclear Tests: Part I (Narrative), 21 April 1978.

67 TNA FCO 66/1311, A. Reeve, Arms Control and Disarmament Department to Mr P.H. Moberly, PS/Mr Hurd and Private Secretary, British Nuclear Test Programme, 12 July 1979 and Carrington to Prime Minister, British Nuclear Test Programme, 17 July 1979.

68 TNA FCO 66/1082, Fred Mulley to Prime Minister, British Nuclear Test Programme, 15 May 1979.

69 TNA FCO 66/1068, Comprehensive Test Ban Negotiations CTB: Date of Cessation of Testing, UK CTB Delegation, Geneva, 20 July 1978.

70 TNA DEFE 66/1082, G.G.H. Walden to Mr Mallaby, British Nuclear Test Programme, 17 May 1978.

71 TNA FCO 66/1069, B.M. Norbury, Head of DS 11, MOD to C.L.G. Mallaby, Head of Arms Control and Disarmament Department, FCO, 8 August 1978.

72 TNA DEFE 19/242, G.G.H. Walden, Foreign and Commonwealth Office to Bryan Cartledge, 10 Downing Street, Comprehensive Test Ban, 22 June 1978.

73 CAB 164/1564, Enclosure 1 to CTB Despatch of 28 July 1980, CTB Negotiations – State of Play on Main Issues, July 1980.

74 TNA CAB 130/1072, GEN 63 (79) 1, Cabinet Official Group on International Aspects of Nuclear Defence Policy, Comprehensive Test Ban: National Seismic Stations in the United Kingdom and Dependent Territories, Note by Officials, 19 January 1979.

75 TNA CAB 164/1564, United Kingdom Delegation to the Comprehensive Test Ban Negotiations, Geneva, Nuclear Test Ban Negotiations 1977–1980: Part II – What Now? 28 July 1980.

76 TNA FCO 46/2286, FCO Telno 11, CTB: Discussions with US Officials, 11 January 1980.

77 'Arms Control and the 1980 Election – Presidential Candidates', *Arms Control Today*, Vol. 10, No. 5, May 1980.

78 For a detailed account on the origins of the 'dual-track' decision see Strobe Talbott, *Deadly Gambits: The Reagan Administration and the Stalemate in Nuclear Arms Control*, London, MacMillan, 1985. For an insight into Soviet deployment of the SS-20 see Jonathan Haslam, *The Soviet Union and the Politics of Nuclear Weapons in Europe, 1969–1987: The Problem of the SS-20*, London, Palgrave Macmillan, 1989.

79 TNA CAB 130/1072, Cabinet Official Group on International Aspects of Nuclear Defence Policy, Comprehensive Test Ban: National Seismic Stations, 11 January 1979.

80 TNA FCO 66/1473, J.C. Edmonds to Mr Reeve, CTB Policy Review: Discussion in Cabinet Office, 2 October 1980.

81 TNA CAB 164/1564, Nuclear Test Ban Negotiations 1977–1980 Part I – Where Are We? J.C. Edmonds UK Delegation to the Nuclear Test Ban Negotiations, Geneva to The Right Honourable Lord Carrington, London, 28 July 1980. See also National Security Archive, George Washington University, Washington DC, 'Memorandum of Conversation, Secretary of State (Cyrus Vance) Meeting with UK Secretary of Defense Pym, Ministry of Defence, 22 May 1979,' pages 10–11.

82 TNA PREM 19/213, UKMis Geneva Telno 405 to FCO, Comprehensive Test Ban: US Attitudes, 17 August 1979.

83 TNA CAB 130/1072, GEN 63 (79) 1st Meeting, Cabinet Official Group on International Aspects of Nuclear Defence Policy, Comprehensive Test Ban: National Seismic Stations, 11 January 1979.

84 The Group was chaired by a Cabinet Office official, but its membership was drawn from senior officials from the FCO, MOD and Treasury. In the British civil and diplomatic service a submission to Ministers is a written document typically containing one or more recommendations, a description of the problem and the background to the proposed course of action on key high level decisions where a Ministerial decision, or guidance is required. It also shows which key departments and departments of state agree with the recommendations.

85 TNA CAB 130/1072, GEN 63 (79) 2, Cabinet Official Group on International Aspects of Nuclear Defence Policy, Comprehensive Test Ban: National Seismic Stations in the United Kingdom and Dependent Territories, 23 January 1979.

86 See John Shepherd, *Crisis? What Crisis? The Callaghan Government and the British 'Winter of Discontent'*, Manchester, Manchester University Press, 2015.

87 An array station is arranged in a regular geometric pattern (e.g. cruciform) with separate seismometers spread out over very large areas. New seismic arrays built by the CTBTO's International Monitoring System (IMS) usually have a distribution diameter of three to four kilometres.

88 For a detailed technical description of seismic means of detection see Alan Douglas, Atomic Weapons Establishment, Blacknest, Brimpton, UK, *Forensic Seismology and Nuclear Test Bans*, Cambridge, Cambridge University Press, 2013.

89 TNA CAB 130/1072, GEN 63 (79) 2nd Meeting, Cabinet Official Group on International Aspects of Nuclear Defence Policy, Comprehensive Test Ban: National Seismic Stations in the United Kingdom and Dependent Territories, 22 January 1979.

90 TNA CAB 130/1072, GEN 63 (79) 4th Meeting, Cabinet Official Group on International Aspects of Nuclear Defence Policy, Comprehensive Test Ban: National Seismic Stations in the United Kingdom and Dependent Territories, 22 January 1979.

91 TNA CAB 130/1072, GEN 63 (79) 3, Cabinet Official Group on International Aspects of Nuclear Defence Policy, Note by Officials, Comprehensive Test Ban: National Seismic Stations in the United Kingdom and Dependent Territories, 26 January 1979.

92 For the ins and outs of British policy-making on the NSS issue in 1979, see TNA FCO 66/1280 to TNA 66/1283, Comprehensive Test Ban (CTB): National Seismic Stations, 1979.

93 Hansard, HC Deb, Vol. 960 cc1691–3, 17 January 1979.

94 TNA CAB 130/1072, Cabinet Official Group on International Aspects of Nuclear Defence Policy, GEN 63 (79) 4th Meeting, Comprehensive Test Ban: National Seismic Stations in the United Kingdom and Dependent Territories, 15 March 1979.

95 These were the role that hydrodynamic, hydronuclear or inertial confinement fusion experiments would play in a CTBT and whether there would be any constraints. Such activities would not be verifiable.

96 TNA PREM 19/41, Cabinet Secretary's Incoming Brief for New PM, Comprehensive Test Ban, May 1979.

97 TNA PREM 19/693, R.L.L. Facer to B.G. Cartledge, No. 10 Downing Street, 26 June 1979.

98 TNA PREM 19/693, PMVR (79) 12 Supplement, Economic Summit, Tokyo 28–99 June 1979, Comprehensive Test Ban – UK National Seismic Stations (NSS) Options, 26 May 1979.

99 TNA PREM 19/693, R.L.L. Facer to B.G. Cartledge, No 10 Downing Street, 26 June 1979.

100 TNA PREM 19/693, FCO Telno 1227 to Washington, Comprehensive Test Ban (CTB) Negotiations: National Seismic Stations (NSS) on UK Territory, 19 November 1979.

101 TNA PREM 19/693, UKMis Geneva Telno 345 to FCO, Comprehensive Test Ban Negotiations, From Edmonds, CTB Delegations, 20 July 1979.

102 TNA PREM 19/693, FCO Telno 1227 to Washington, Comprehensive Test Ban (CTB) Negotiations: National Seismic Stations (NSS) on UK Territory, 19 November 1979.

103 TNA PREM 19/693, Extract from Record of Meeting between PM and President Carter, 17 December 1979.

104 TNA PREM 19/693, Washington Telno 4255 to FCO, CTB, 18 December 1979.

105 TNA CAB 164/1664, J.C. Edmonds, Nuclear Test Ban Negotiations, 1977–1980, 8 May 1981.

106 TNA CAB 130/1072, Record of a Meeting held in Conference Room B, Cabinet Office, Comprehensive Test Ban Negotiations, 16 February 1979.

107 TNA CAB 164/1564, United Kingdom Delegation to the Comprehensive Test Ban Negotiations, Geneva, Nuclear Test Ban Negotiations 1977–1980: Part II – What Now?, 28 July 1980.

108 TNA FCO 66/1473, A. Reeve to Mr Callan, Comprehensive Test Ban: Policy Review, 30 April 1980; TNA FCO 66/1473, A. Reeve, Arms Control and Disarmament Department to PS/Mr Hurd, Call by Mr Edmonds: CTB, 2 October 1980.

109 TNA CAB 164/1564, Manuscript minute from L.M. Hastie-Smith to Mr Wade Grey, CTB, 23 September 1979.

110 TNA CAB 130/1158, MISC 1 (81) 1st Meeting, Official Group on International Aspects of Nuclear Defence Policy, Nuclear Test Ban Policy, 19 January 1981.

111 TNA FCO 66/1473, A. Reeve, Arms Control and Disarmament Department, FCO to D.C. Fakley, ACSA (N), MOD, CTB: Policy Review, 23 October 1980.

112 TNA CAB 130/1158, MISC 1 (81) 1st Meeting, Official Group on International Aspects of Nuclear Defence Policy, Nuclear Test Ban Policy, 19 January 1981.

113 TNA CAB 130/1158, MISC 1 (81) 1, Cabinet Official Group on International Aspects of Nuclear Defence Policy, Nuclear Test Ban Policy: Draft Report, Note by the Secretary, 12 January 1981.

114 TNA CAB 164/1664, D.G. Martin, Arms Control and Disarmament Department, FCO to Miss V. J. Bell, DS 17, MOD, UK/US Political Consultations: CTB, and draft points to make, 21 May 1981.

115 See for example TNA FCO 66/1483 and TNA FCO 66/1484, Non-Proliferation Treaty (NPT) Review Conference, Geneva, August–September 1980; 'NPT Review Conference No Declaration', *Nature*, Vol. 287, p. 98, 11 September 1980.

116 TNA CAB 164/1664, D.G. Martin, Arms Control and Disarmament Department, FCO to C.H. O'D. Alexander, Cabinet Office, Comprehensive Test Ban, 24 December 1981.

117 TNA PREM 19/693, Carrington to Prime Minister, Comprehensive Test Ban Treaty, 21 January 1982.

118 TNA CAB 164/1664, Dr Robert Press, Cabinet Office to D. G. Martin, Arms Control and Disarmament Department, FCO, Comprehensive Test Ban, 12 January 1982.

119 TNA PREM 19/693, Carrington to Prime Minister, Comprehensive Test Ban Treaty, 21 January 1982.

120 TNA PREM 19/693, A.J. Coles 10 Downing Street to Bryan Fall, FCO, Comprehensive Test Ban Treaty, 25 January 1982.

121 TNA FCO 66/1586, D.H. Gillmore to PS/Mr Hurd, CTB, 4 February 1982.

122 TNA FCO 66/1586, Washington Telno 688 to FCO, My Telno 633: CTB, 2 March 1982.

123 TNA FCO 66/1587, Washington Telno 1324 to FCO, UKDis Geneva Telno 26: CTB in the CD, 16 April 1982.

124 TNA PREM 19/693, F.N. Richards, FCO to A.J. Coles, 10 Downing Street, Comprehensive Test Ban (CTB), 7 May 1982.

125 William Epstein, 'A Disastrous Decision', *Arms Control Today*, Vol. 12, No. 8, September 1982.

126 Hansard Written Answers, Vol. 427, cc820-1WA, 22 February 1982.

127 TNA 66/1588, J.S. Chick, Arms Control and Disarmament Department to Mr Gillmore and PS Lord Belsted, Parliamentary Question: Lord Jenkins, CTB, Background Note, 11 October 1982.

128 Hansard, HL Deb, vol. 418 c511, 16 March 1981.

129 TNA FCO 66/1587, UKDis Geneva Telno 27 to FCO, Washington Telno 1324 to FCO: CTB in the CD, 19 April 1982; TNA FCO 66/1587, UKDis Geneva Telno 28 to FCO, Our Telno 27: CTB in the CD, 21 April 1978.

130 Lawrence Freedman, *The Official History of the Falklands Campaign, Vol. II, War and Diplomacy*, London, Routledge, 2007, pp. 13–27.

131 TNA PREM 19/693, R.B. Bone, Private Secretary, FCO to A.J. Coles, 10 Downing Street, Arms Control and Disarmament, 29 November, 1982. Disarmament and Confidence-Building Measures which the UK has proposed or is supporting: 1979–1982. Hansard, House of Lords Debates, vol. 434 columns 819–21, 13 October 1982; Hansard House of Lords Debates, Vol. 434 column 813 Written Answers, 12 October 1982. For a good summary of the purpose and history of the Group of Scientific Experts see Ola Dahlman, Frode Ringdal, Jenifer Mackby, and Sven Mykkeltveit, 'The Inside Story of the Group of Scientific Experts and its Key Role in Developing the CTBT Verification Regime', *The Nonproliferation Review*, pp. 181–200, 3 June 2020.

132 TNA 66/1589, J.S. Chick, Arms Control and Disarmament Department to Mr Gillmore and Parliamentary Unit, Oral Parliamentary Question: Mr Frank Allaun: Mr Gromyko's Proposal for a CTB, Notes for Supplementaries, 17 November 1982. For academic views, see Lyn R. Sykes and Jack F. Evernden, 'The Verification of a Comprehensive Test Ban Treaty', *Scientific American*, Vol. 247, No. 4, 1982.

133 TNA DEFE 19/242, Percy Craddock, Head of UK Delegation to the CTB Negotiations, FCO to David Owen, Negotiations for a Comprehensive Ban on Nuclear Tests: Part II (Comment), 21 April 1978.

134 Christopher Mallaby, *Living The Cold War: Memoirs of a British Diplomat*, Stroud, Amberley Publishing, 2017, electronic version location 2027.

3 British Nuclear Tests 1974–1982 and Test Ban Pressures

Introduction and context

After a gap of nine years, during which the effectiveness of AWRE as a warhead design laboratory had deteriorated considerably, the UK resumed underground nuclear testing on 23 May 1974.[1] A further test was conducted on 26 August 1976. In this period (1974–1976), there were no negotiations on a CTBT underway or any in the offing. This would change in 1977. As noted in Chapter 2, the UK joined the US and USSR in what became known as the tripartite test ban talks in Geneva. The prospect of a test ban treaty in force posed a threat to the maintenance and modernisation of the UK's strategic and tactical nuclear weapons programmes until 1982. However, the threat was much diminished between 1980 and 1982, when the negotiations were suspended sine die in 1980 as we saw in the last chapter. So for a short period of roughly three years, the UK, as it had in 1954–1958, faced competing objectives: support for a test ban treaty whilst also trying to preserve the credibility of its nuclear weapons capability – a capability which would require further underground tests if there were to be safe and reliable successors to Chevaline and the WE177s. There is little technical detail publicly available about the purpose of the individual nuclear tests conducted by the UK in this period. Annex 1 provides a list of the UK tests conducted between 1974 and 1986 and summarises what is known and can be deduced from papers in The National Archives. In all but four cases the actual yield of the tests remains classified. The rationale for this is not entirely clear given that the yield of all the UK's atmospheric tests in the 1950s has been made public, and that all but the proof tests of the Trident warhead were of devices that were not intended to enter the stockpile. We also need to note that the preparation of an underground nuclear test at Nevada took some considerable time from the design and fabrication of the test device, to the formal UK request to the US, then for submission to the President for final authorisation, through to drilling the borehole, instrumenting the device, lowering the test rack containing the assembled device and backfilling and stemming the borehole prior to firing. This explains why the UK needed about 12 months between seeking US Presidential approval for a test and the desired firing date.[2] Nuclear tests were thus not spur-of-the-moment decisions.

DOI: 10.4324/9781003375708-3

In this context, we should also note that both the primary (the fission trigger) and secondary (thermonuclear) components of a nuclear weapon had to be tested before being built into a production warhead intended for operational deployment. However, the secondary, unlike the primary, could be tested at less – sometimes much less – than its full yield, the penalty for doing so being an increase in the uncertainty in the estimate of the full yield. Had a threshold test ban treaty been set at not less than 30 kilotons that would have enabled the development of warheads for Trident and for the foreseen theatre nuclear weapon requirements subject to about 20% uncertainty in achievable operational weapon yields.[3] Stockpile maintenance could be sustained with a threshold at 5 kilotons. These observations are particularly important to keep in mind as we look at the development of the future Trident warhead and the number of tests required to acquire the necessary design information and validate the production design.

We should also consider the context of UK nuclear testing. Aldermaston and the MOD saw regular testing as primarily part of a necessary research and development programme that was essential if the UK were to remain in the nuclear business.[4] In mid-1973 an essential component of such a programme included the development and design of progressively harder warheads against improving Soviet anti-ballistic defences and possibly an increase in their yield in order to diminish the number of penetrations necessary by in-coming warheads needed to provide the level of destruction deemed necessary by the UK's deterrent criteria as established by the Joint Intelligence Committee.[5] Such a programme would, however, be totally ineffective without the ability to have occasional nuclear tests of the concepts produced by AWRE's theoretical work. Furthermore, against the background of a possible comprehensive test ban treaty, or a limitation of underground tests, MOD officials thought it would be most unwise to postpone AWRE's warhead research programme whatever the timescale of future strategic successor systems. Essentially the more data gathered from underground tests the better would be AWRE designers' knowledge and capabilities. Such an assessment, which even though it was made in mid-1973, is still significant for two reasons: first a CTBT was not a realistic possibility at that time, and secondly it provides a clear statement on the centrality of periodic nuclear testing to maintain capabilities and options. This latter point is a recurring and central theme in the story of British nuclear weapons and the test ban in the period covered by this book.

Britain's ability to keep its nuclear weapons in a safe and reliable condition had up until the mid-1970s been maintained through a programme of nuclear warhead R&D. However, with the successful completion of much of the development work for the new Chevaline warhead by May 1976, Aldermaston and the MOD were presented with the problem of maintaining its research programme without knowing whether the UK was likely to need any future nuclear warheads – strategic or tactical, or the nature of the operational requirements for them. For this reason, the MOD's Chief Scientific Adviser Sir Herman Bondi asked Sir John Kendrew FRS in 1974, assisted by Sir Willian Hawthorne FRS and a Professor Tompkin to review a possible programme of research outlined by the MOD's Deputy Chief Scientific Adviser (Programmes and Nuclear) and the Director of AWRE, and report on the most modest but effective warhead research programme.[6] Their report, which was given a very limited distribution – a not uncommon occurrence on nuclear weapon matters,[7] defined some programme options that were directed principally towards obtaining a basic understanding of the design features of small and lighter nuclear warheads for both tactical and strategic roles, and in addition strongly recommended work on the application of very high power lasers to investigate thermonuclear processes of direct relevance to nuclear warhead work. One of the reasons for this interest in lasers was that they would, it was thought at that time, play a big part in nuclear warhead R&D if and when a comprehensive test ban treaty came into force. For this reason, Kendrew and his two colleagues thought that it would be important for the UK to understand what was and what was not possible in this field. Moreover, a small but concentrated laser programme would provide sufficient value to give AWRE access to the results of the large US programme. Bondi strongly endorsed the Kendrew report's key recommendation and argued that the only sensible course for the UK would be to embark on a programme to learn what would be required to design smaller and lighter warheads. Bondi recognised, however, that politically this would be difficult for a Labour Party Secretary of State for Defence to agree to a programme that entailed periodic nuclear testing, or implicitly on a decision that presumed that a new strategic or tactical warhead would be the end result of such a testing programme. However, if the UK was to at least retain the option of acquiring future warheads, then there was no viable alternative, but more significantly AWRE could not be confident that it would be able to maintain the safety and reliability of the UK's existing nuclear weapons for very long without such a programme. The formal requirements of the US nuclear test programme, together with the rundown of Chevaline work, meant that the UK had to decide by November 1976 whether to embark upon a new programme that would require its first nuclear test at about the end of 1977. However, given the need to book a slot in the US programme in good time, then the UK would need to make a provisional commitment at the end of May 1976 in the annual UK–US nuclear stock-taking meeting held under the auspices of the 1958 Mutual Defence Agreement. Ministers, however, would have to approve a specific test or programme of tests, and a final decision would not be likely until the end of 1976. As we will see, approval was eventually given for a new round of tests, but not in 1977 as the Defence Secretary required much more detail on the proposed programme, its purpose, staffing and costs before reaching a decision.[8]

In early September 1976, Sir Hermann Bondi, in response to the Secretary of State's request for more detail, stressed once again that unless the nuclear warhead research programme was authorised that autumn, the US would start to have serious doubts about the UK's will to continue as active partners in the nuclear weapons field. Cooperation would suffer accordingly.[9] This line of reasoning appears repeatedly in this period – unless the minimum R&D efforts were maintained, then the UK would risk going out of the nuclear business altogether. It seems as if it were posited as a rhetorical question to Ministers – do you really want to be the ones to take the UK out of the nuclear weapons business? Victor Macklen (DCA PN, MOD – the central nuclear policy and scientific post in MOD) had prepared a detailed paper setting out the options and requirements for a future research programme. AWRE was considering two possible routes for the development of a small light warhead that might be needed at some time in the future.[10] One of these was an extrapolation from an existing warhead design, the other was a new technique which the UK knew that the Americans and French had explored and tested. The objective for either route would be to produce a warhead that would be suitable if a requirement were ever to arise for a strategic system to counter the deployment of a terminal ballistic missile defence, or if there were a need to extend the operational range of the Polaris missile system. These attributes would also be relevant for future tactical weapons. Another line of research could lead to the reduction in the amount of plutonium required for tactical weapons. As already noted, the absence of UK testing between 1965 and 1974 had led to a considerable deterioration in AWRE's effectiveness. And this experience was to have a clear and recurring impact on MOD nuclear policy in our period. Such was the decline, that the UK's lack of originality and generally slow progress in 1965–1974 prompted some severe criticisms from the US in 1972. Without a test element, AWRE and MOD feared that the US would not continue to support the detailed exchange of nuclear technology upon which the UK's nuclear capability depended.[11] However, a programme of testing roughly once every 18 months would suffice to prevent a painful recurrence of this embarrassing experience. The Defence Secretary duly noted Macklen's recommendations and planned to consult certain Ministers and the PM before deciding on what course of action to take.[12] A key question to consider here is whether all of this pressure from Bondi and Macklen represented an element of the scientists and officials prejudging future political decisions on a replacement for Polaris by insisting that some work had to continue if Ministers wanted the UK to remain a nuclear weapons state; or whether this was solely an attempt by AWRE and the MOD to preserve a basic capability in nuclear defence science at the minimum level necessary for the UK to remain in the nuclear business. Keeping future options open was always a feature of good civil service contingency planning.

Tests 1974–1978

The inevitable consequences of the decision to modernise Polaris – the Chevaline programme – in order to improve its ability to defeat Soviet ABM defences were that new nuclear tests were unavoidable. Changes to the warhead could not be accommodated without prior testing. A new hardened warhead for Chevaline was required in order to protect the physics package against neutrons generated by an ABM explosion in outer space and for the miniaturised hardened warhead then being developed to preserve the range of the new Chevaline front end.[13] The Fallon test, originally planned for the end of 1973,[14] of 23 May 1974 was thus held to prove the 'Harriet' warhead design. In the absence of a large scale UK test programme warhead reliability would have to be assured by a thorough understanding of the effects of hardware variations. Those which could only be observed in a nuclear test were built into a second test round in 1976, of which more later.[15] In addition to a full warhead test, AWRE also had space in US underground flux tests to expose key components to radiation and neutron levels that would likely be encountered in space if a re-entry body faced an ABM nuclear explosion. A test code named R6 in early 1976 was planned to conclude the bulk of the flux testing of Chevaline components.[16]

Despite the success of the Fallon test of a radically new design, it had not provided sufficient information for AWRE designers to be able to give a guarantee to keep this new design in a reliable and safe condition over the expected 10–15 years of its service life. Therefore, in order to acquire this information, a further one or two tests of the basic design were needed. Although the test in 1974 had been successful in demonstrating the basic design yield of the hardened warhead for Chevaline and that an intrinsically safe warhead could be manufactured,[17] Aldermaston needed a further test to provide a greater depth of knowledge of the design, which would then able its scientists and engineers to cope confidently with any in-service problems over its years of service life. As a bonus, these additional tests might also enable the UK to economise on the manufacture of the warhead, especially in its use of plutonium.[18] Plutonium was always expensive and the UK programme frequently had a shortage of it. This new test was intended to assist in a fundamental re-examination of the whole system weight/range, to clarify the 'exchange ratio',[19] and as a prototype warhead to test against future defensive ABM warheads. Final approval from the Prime Minister was sought some six weeks before the second test in case he decided there should be some prior discussion in Cabinet.[20] Callaghan's predecessor, Harold Wilson, had come under fire from some of his Ministers in the Cabinet after the May 1974 test.[21] The left-wing of the Labour Party was opposed to any replacement of Polaris, and consequently always likely to give the PM grief over the issue. However, the Prime Minister took his time in issuing formal final approval for the test to take place, a fact that became somewhat acute when the US informed the UK that arrangements had been made to fire the test on 26 August and needed confirmation to proceed by 12 August. In fact, the UK would need to confirm sooner as the US had in fact emplaced the test device down a 2,000-foot shaft on 28 July and would likely not be willing, or even able to bring it back up again. If the UK could not confirm the firing of the test, then the slot would be lost and could not then be re-scheduled until well into the US test programme for 1977. Moreover, the UK would also forfeit a significant part of the £3.1 million that the test was costing the MOD.[22] Officials feared that a delay of such length would have a serious effect on the Chevaline programme, but the Prime Minister was in no rush to take a decision in advance of a meeting on the wider Chevaline programme's status and progress. A decision was potentially pending on whether to continue with the programme given some of the engineering and cost problems that it had faced. Unfortunately, for MOD officials, this meeting of key Ministers had been deferred until 29 July 1976.[23] Fortunately, it proved possible, without any awkward questions being asked by the Americans, to delay events so that nothing irrecoverable would take place at the Nevada Test Site (NTS) before 30 July.[24] MOD officials had made clear to Ministers that the August test was an essential feature of establishing the safety of the Chevaline warhead over its prospective life; without this test, AWRE could not guarantee to keep the warhead in service over 12–15 years. 'Safety' is the magic word no doubt stressed to ensure that Ministers took the right decision. AWRE also needed further information about the possibility of economising in the manufacture of the warhead, something which could only be acquired from this test. Macklen drew the Defence Secretary's attention to the fact that if Ministers did decide to cancel the Chevaline programme, then there would still be value in proceeding with this test if the UK still wanted to retain a nuclear R&D capability. Aldermaston's re-designed Chevaline warhead was based upon a concept new to the UK of initiating the primary, and such a concept would be the basis for any likely future nuclear warhead for either tactical or strategic use. The August test was thus also designed to extend AWRE's understanding of this new concept; furthermore, AWRE had also added various experiments to the device to extend the range of information obtained still further in an attempt to get two for the price of one. Since the test device could not be transferred back to the UK under US law, and given the added potential prospect of a CTBT, all this added up to the high desirability of proceeding with the test as planned unless Ministers wished to abandon the UK's capability to design nuclear warheads.[25] Macklen was clearly trying to frighten Ministers into taking what, for him, would be the only sensible, obvious and pragmatic decision, that is, fire the test device. Assembly of British nuclear test devices in the US could only be done under US law if the device was in US custody. Section 92 of the 1954 US Atomic Energy Act prohibited the export of atomic weapons from the US to any third party. In practice, the MOD and US Energy Research and Development Administration (the lead US agency on nuclear weapons testing at that particular time) had deliberately never considered the strict legal position on the grounds that the UK device once exploded did not exist and for the few days it did exist it was in US custody.[26]

At the Prime Minister's meeting with the Defence and Foreign Secretaries and Chancellor of Exchequer on 29 July 1976, Ministers agreed that the Chevaline programme should continue and that the August test should proceed as planned. The PM would inform Cabinet of the planned test before the parliamentary recess.[27] There was no dissent in Cabinet from Callaghan's statement that an underground test should be held in Nevada on or about 26 August when it met on 3 August.[28] The way was thus clear for the test to proceed. Banon was fired at Nevada on 26 August 1976. There were no UK nuclear tests in 1977.

From the results obtained in these two tests, AWRE was able to develop an improved warhead for Polaris for service use. However, Aldermaston's weapons' designers believed it would be possible to make this warhead more effective operationally by introducing modifications (primarily by making it physically smaller). These modifications needed to be proved by a further nuclear test scheduled for the spring of 1978, which explained MOD's worries over a test ban treaty or moratorium on testing as either would prevent the test from taking place. If such a prohibition on tests were in place, this would amongst other things deprive the UK of a further understanding of this smaller type of warhead and of an enhanced ability to maintain it in service without recourse to further testing.[29]

A nuclear testing R&D programme as recommended by Bondi, when eventually agreed by Ministers sometime between October and February 1977,[30] had as its prime focus the design of smaller and lighter warheads that could be useful if there were to be a subsequent decision on acquiring a successor to Chevaline or replacement for the WE177s tactical bombs. In late February 1977 Macklen informed the Defence Secretary that since their previous discussions on a nuclear testing programme, a second possible design for the UK's next nuclear test had been discussed with the US, which was also aimed at producing a lighter warhead.[31] This would be a shorter-term modification of the new Chevaline warhead design, which could prove to be the cheapest way of ensuring that the UK's improved Polaris force had the longest effective operating range, which in turn would give the Resolution class submarines greater sea-room for patrol. Both designs would be of value in increasing AWRE's design capability and maintaining close UK-US exchanges of nuclear information. Given an early decision, either device could be ready for testing in autumn 1978. In view of the Prime Minister's planned trip to Washington to see President Carter on 10 March 1977 during which Callaghan planned to discuss nuclear arms control, both the FCO and Cabinet Office advised the MOD that there was some urgency in submitting to the PM the request for the start of preparations for a further British nuclear test. There was, in Macklen's view, no direct inconsistency whatever between attempting to move towards a CTBT, which was necessary in the long term and making prudent preparations to maintain the UK's nuclear capability in the nearer future. Macklen wanted to book a slot in the US test programme for autumn 1978. If for any reason, such as proposals for an early CTBT looked bright, the test could still be postponed, or cancelled right up to the last few weeks or days. This was Macklen no doubt making sure that CTBT prospects would not prevent the necessary preparatory work given the long lead times involved in arranging an underground test at Nevada. The Defence Secretary, therefore, sought the PM's agreement on 25 February 1977 to proceed with the preparations for a test in autumn 1978.[32] Callaghan agreed but instructed the Defence Secretary to seek final approval nearer the time in light of the consideration of his further report on the progress of the Chevaline programme.[33] The US planned to provide more information to the UK, which would speed up the UK's timescale, but even if the Americans provided drawings of a US-tested warhead, the UK would still need to test its own production models, probably only once.[34] Meanwhile, the Foreign Secretary asked whether it would be possible to bring forward the planned date of October 1978 for the next test.[35] Dr Owen was sufficiently worried that progress in CTBT negotiations might move rapidly as to make an October 1978 date politically embarrassing for HMG.[36] Fortunately, from both a British and US point of view, it was practicable to bring the date of the next test forward to March 1978. The financial consequences of bringing the date forward were relatively small and confined to the UK. Although the US costs were greater than for the 1974 test because of inflation and the higher planned yield of the 1978 test, there was sufficient in the costing estimates to cover the earlier phasing of the October test. AWRE and the US laboratories had also reviewed the technical risks of the earlier date, and subject to some deletions of some of the peripheral research add-on experiments, all US labs agreed that the earlier test date would not prejudice the chance of success of the British design, which was rated as good. In light of this Prime Minister authorised the proposed advance of the UK nuclear tests, but entered certain reservations about the new date if it became clear that the UK, US and USSR were likely to reach an agreement to cease testing for a limited period of 18 months.[37] As ever optics mattered. Callaghan commented that it might make the UK look shabby if a test were conducted some months before a testing moratorium began. It would be important, therefore, to find out whether either the US or the Soviet Union will continue testing up to the last minute.[38]

However, the CTBT negotiations were by spring 1978 moving too slowly to make very real the prospect of a moratorium on testing preventing the March UK test; and in any case, the test itself was not essential to the completion of the Chevaline programme.[39] Apart from a unilateral US political gesture suspending tests in the anticipation of an acceptable treaty, MOD officials saw no reason to suppose that this test would not proceed as planned.[40] If successful, the test would allow the UK to produce a modified warhead with important operational benefits for Chevaline as already noted. Beyond this, MOD and AWRE looked to see what further designs could be tested on the assumption that a test ban treaty would commence not earlier than the end of 1978. Even with a crash programme it was not going to be possible for AWRE to test designs incorporating a super safe chemical explosive, which might do much to reduce both operational vulnerability and terrorist threats to hijacked weapons. However, combining some of the principles being investigated in the planned March 1978 test with a research project would produce a device for test in September 1979. This would be a smaller diameter light weight warhead design aimed at an eventual successor strategic system whether it be a cruise or ballistic missile. As the US authorities were now urgently considering their own crash programme of underground tests, the UK would need to secure one slot in this programme, and given the timescale, would need to approach the Americans before the end of November 1977. Accordingly, the officials needed the Secretary of State for Defence to consult the PM to authorise the approach. Tellingly, officials noted that if this were to be UK's last chance to test a British device, it should not be lightly abandoned. Moreover, if for political reasons, the test was cancelled before firing, the fact that the UK had intended to have a test would provide it with additional leverage in seeking advanced and proven warhead designs from the US. It seems that the pressure to book and preserve slots in the US testing schedule was used by MOD officials to secure Ministerial agreement.

There is an interesting footnote to the plans for the March 1978 test. Normally the UK and US made no prior announcement of any UK test; however, on this occasion, the planned yield of the device was such that US policy required a prior announcement for public safety since the ground motion for a test of this size might be felt outside the Nevada Test Site. US authorities in such circumstances warned the local population to avoid being on ladders, or on the edge of high buildings at the time of the explosion. The pre-shot announcement would be in the standard US form and make no mention of the UK. AWRE could not reduce the planned yield of the test and its purpose would be invalidated if it were further reduced.[41] The MOD did note, however, that the yield was still 'well below the agreed threshold of 150 kilotons'.[42]

Modifications to the previous design tested in 1976 were eventually further examined at the Cresset (Fondutta) test fired on 11 April 1978, the original planning date had been in March.[43] As far as we can see from available public sources at The National Archives, this test was used to examine further the elements explored in the Banon test. In addition, the April 1978 test was used to prove the performance of a 'reduced weight version of the Chevaline warhead',[44] as well as seeking to improve AWRE's knowledge of the Chevaline design.[45] Indeed a primary reason to proceed with the underground test was to provide a light weight warhead for Chevaline.[46] Apparently, it also included 'a significant design change',[47] but it is not at all clear from the files at Kew what this change entailed. However, we can be sure that a new light weight warhead for Chevaline was tested successfully at Fondutta, using the Findhorn/Firstrate primary (Findhorn was the UK codename for the device tested at Fondutta),[48] and part of the Battler (see Chapter 5 for a more detailed discussion of the Battler project) modification to the Chevaline payload. Fondutta demonstrated the UK's ability to put a lighter warhead into later Chevaline missiles, if it was so decided as this would give an increased range for the missiles and thus more sea room for the Resolution class submarines to operate in.[49] We also can see that the warhead for Chevaline had been 'satisfactorily tested at nuclear yield'.[50] Since the US had signed the Threshold Test Ban Treaty, then the yield could have been no more than 150 kilotons and probably much smaller in order to ensure that the maximum credible yield of a device would not run a risk of breaching this limit, the actual yield would have been closer to 100 kilotons. See the Annex for further discussion of this point.

Following this successful test in April 1978, the MOD commissioned studies at AWRE and in industry to investigate the design, performance and cost implications of incorporating the new lightweight warhead into Chevaline.[51] Initial estimates suggested that there would be no major problems and that an increase in range of about 64 nautical miles would be obtained by using the smaller warhead, although there would be some reduction in the warhead yield. This suggests perhaps that this design used less fissile material. AWRE and the MOD thought that it should be possible to introduce the modified warhead into service at the third outload of missiles on the Resolution class SSBNs in 1983.

The Defence Secretary now proposed a third UK test for 1978, which was additional to the two already provisionally agreed by Ministers. AWRE's precise objective for a third test had not yet been set, nor was the test of a particular operational need, but, and this is the significant point, it would add to the store of knowledge on which the UK would need to rely on in the event of a CTBT as it would provide a stock of proven technical knowledge that would have to serve the UK for years to come in a world without testing.[52] It was also possible that the test would be designed to enable the MOD to explore the principles of small and hard warheads.[53] If this test were successful, such an outcome would enable the UK to produce new warheads without further testing if it were decided to replace part of the tactical nuclear weapon stockpile, or produce warheads suitable for cruise missiles or to further improve Polaris in the event of developments in Soviet ABM capabilities. From the FCO's perspective, three UK tests in a single year would represent a considerable increase over the average rate of testing. As already noted, there were no UK tests between 1965 and 1974 or in 1977. Officials also observed that were a CTBT to prove likely, there would be significant pressure for an immediate moratorium on further nuclear testing until the treaty entered into force. If that were to be agreed by the US and Soviet Union, then the UK would have to follow suit to the detriment of its own testing plans. MOD wanted to seek a provisional slot in the US programme for December 1978. The successful April test had, moreover, led to the release of more up-to-date US technical information to UK weapons designers at AWRE, and if the September and the possible December test were also successful, then this would likely prompt the release of further US data, which would be of immense value to the UK. Continued UK testing, especially of innovative designs, provided the keys to the cupboard containing US warhead design data, which was one of the main reasons why AWRE and MOD were so keen to keep the UK's testing hand in. Furthermore, all this would clear the way to fruitful UK/US collaboration on stockpile maintenance problems in a CTBT.[54] A provisional commitment to a UK test in December would require funding of $10 million for contingency preparations. However, the Prime Minister took the view, given all the considerations – the Foreign Secretary also objected to a third UK test in 1978, noting too that the CTBT negotiations were an important factor in determining the timing of UK tests,[55] that the September nuclear test should not proceed. Callaghan did agree that, subject to final approval, preparations for a test in December test could continue.[56] The MOD simply postponed the September test until December.[57]

The second test for 1978 was based on the same design principle as the test conducted on 11 April 1978, and which, if successful, would provide much-needed information for future stockpile reliability assessments. The device for this test would be physically smaller than the one tested in April and thus provide a more flexible choice for any future options for successor nuclear systems. The test was eventually fired in November 1978, brought forward from the initial date of December.[58] Preliminary results indicated that everything worked as expected.[59] In view of the close comparison of the measured and predicted yields, AWRE regarded this as a successful event. Although Quargel was largely a repeat of the technology tested at Fondutta,[60] it did explore further the new warhead principles of Fondutta.[61] Quargel was also a lower yield device based on the same design principle,[62] and physically smaller than the one tested at Fondutta,[63] which meant that its reduced size might be suitable for a pointed re-entry body (RB), rather than the blunt-nosed slower Chevaline RB. As such this test was the UK's first attempt at a device which would allow a high-speed re-entry vehicle for a ballistic missile.[64] It seems, therefore, that Quargel was especially important in the context of a future strategic warhead. Denis Fakley, Director Defence Science 6 in the MOD's Assistant Chief Scientific Advisor (Nuclear) staff subsequently noted, 'We are also very keen to get the Quargel test under our belt because it should put us in a position to respond to the full requirement for the warhead for any successor system that might be chosen'.[65]

However, before AWRE could bring such an experimental device into safe and reliable warheads for service use, it still required further information from its test planned for July 1979.[66] If that test were to be successful it would begin to open possibilities for increasing the yields of the devices and would certainly open the door to far more information exchange with the US on their devices of a similar nature. Moreover, it would also provide much-needed background information for stockpile reliability purposes.[67] These experimental devices included other side experiments, which allowed AWRE to gain information of great value in relation to other warhead designs. As ever the UK had to maximise the amount of information that it derived from every underground nuclear test given their expense and their domestic political sensitivities for a Labour government, especially in the context of CTBT negotiations, or a prospective or actual testing moratorium. AWRE scientists simply did not have the option of blasting away to their heart's content in the way that their US, Soviet and even French counterparts managed.

Macklen informed the Secretary of State on 6 June 1978 that as for the timing of a further UK test at Nevada, the US position was that they could certainly provide a slot for a UK shot in the period between February and July 1979. However, the range of dates depended upon the scientific measurements that AWRE wanted to make. Essentially, the more complicated these were, the later the test would need to be given the extent and nature of the diagnostic equipment, cabling and recording trailers that would be required. However, the later the test, the greater the risk that it might be prevented by a CTBT, but on the other hand, if there were to be an eventual test ban, this placed a premium on obtaining the maximum amount of scientific information beforehand.[68] However, the MOD thought that there was a reasonable chance that CTBT negotiations would not have proceeded far enough to impede this test.[69] They were right.

Tests 1979–1982

Although a CTBT was unlikely to enter into force before July 1979, the Foreign Secretary felt that the Geneva tripartite negotiations should be concluded as quickly as possible – preferably by the end of 1978 and that by July there would likely be a US/UK/USSR treaty under discussion with the non-nuclear weapons states.[70] There were also pressures for a testing moratorium until a treaty entered into force; however, the US had not decided whether to observe such a moratorium; and if it did so, no view was taken on at what stage of the CTBT process it should be put in place. Despite this, the Foreign Secretary noted that a moratorium commencing from the time of treaty signature was thought quite possible in some quarters of the US Administration – although it would have been very doubtful that either the Department of Defense or Joint Chiefs held such a view,[71] so the Foreign Secretary must have been thinking of State Department or ACDA. In these circumstances, Dr Owen believed that Ministers needed more information on the need for the July 1979 test and its costs before preparations should proceed. In response, the Defence Secretary argued that the proposed device was of exceptional technical importance in maintaining all the UK's options for future deterrent systems. MOD officials worried that any delay on the UK part would result in the loss of the slot earmarked for the British test.[72] Echoes of the arguments used in previous years are evident here. Future strategic delivery systems and the defence environment that they would need to penetrate would impose severe size and weight constraints on warhead designers. AWRE's proposed test in July 1979 was intended to explore the technology of 'very small triggers in a new area', though it is not clear from declassified papers what was meant by '*in a new area*'. This would be only the second test that the UK had conducted relevant to small hard warheads and the related information derived from such a test would be of great value in the future, especially if a test ban treaty were agreed.[73] AWRE's design (Dicel) was of a new concept, which was much more advanced than Quargel.[74] Nessel, the US codename for the test was described as being of exceptional technical importance for future nuclear weapons. Furthermore, the US had agreed to carry out the most extensive measurements including some parameters that the UK had not been able to measure before. These measurements would give AWRE designers a clearer picture of the crucial stages of the implosion of the very small fission trigger that would light the thermonuclear fuel in the secondary.[75] A better understanding of detonation physics was, moreover, relevant to stockpile maintenance problems of all then-modern nuclear warheads.[76] Such information would provide possibly crucial information relevant to keeping UK options open for future strategic deterrent systems, stretching beyond the end of the 20th century.[77] Los Alamos and Lawrence Livermore weapons labs were both especially interested in this UK nuclear test and thus very keen to see it conducted.[78] This great interest underlined to Macklen the value of this test in enlarging the scope of existing close UK-US cooperation on warhead design. Such close collaboration would become doubly important in maintaining UK design competence under a possible CTBT.[79] AWRE assumed that future warheads would need to be constrained in size and weight.[80] Presumably, they assumed this because a future UK strategic warhead would be required for a MIRVed ballistic missile system where smaller warheads would be needed to fit inside the cone-shaped high-speed re-entry bodies fitted onto the bus at the business end of the missile. The novel feature of the Dicel design was its very small nuclear trigger.[81] Dicel had been fully discussed with US experts who recognised its advanced features and as a result offered to apply their latest diagnostic techniques to the test.[82] This development would greatly broaden the UK's capability to design small 'packagable' nuclear warheads, which could be applied to a wide range of theatre or strategic delivery systems.[83] Dicel would also begin to open the possibilities for increasing the yields of warhead designs suitable for high-speed re-entry vehicles.[84] Interestingly enough after the Thatcher government came to power in May 1979, Nessel was described to Ministers as related to 'proving a British warhead design for a successor system'. This is perhaps the first suggestion in the archives that the eventual warhead for Trident can be directly traced to Nessel even though the test was approved under a Labour government and almost a full year before the Thatcher government announced its decision on 15 July 1980 to Parliament to buy Trident C4 SLBMs from the US.[85]

MOD officials also noted that if the UK were unable to test in July 1979, any expenditures would not necessarily be nugatory. Any facilities at the Nevada Test Site if not used because of an early test ban would be reserved for UK use if testing were resumed at low yield for stockpile maintenance after three years following the expiration of the treaty. AWRE could not guarantee that a need for a British nuclear test for stockpile maintenance purpose would not arise, and in that case, the existence of an already drilled borehole could save the UK six months in preparation time. MOD officials also worried that the UK's ability to design and produce modernised and improved theatre nuclear weapons – a WE177 replacement – would be seriously curtailed by the early entry into force of a CTBT, or by agreement to any prior moratorium on testing.[86] Across the road in King Charles Street their FCO counterparts believed that agreement of a treaty in the tripartite negotiations by the end of 1978 was achievable, or at least by not later than the spring of 1979.[87] However, the Soviet position was that there should be no multilateral phase of negotiation in the CCD – the UK and US preference, so there was a risk that by summer 1979 there might be a treaty awaiting ratification in the UK, US and USSR – hardly a good time to be testing in FCO thinking. The Foreign Secretary himself was also wary of committing to a nuclear test at this time for much the same reason.[88] Dr Owen went on to advise the Prime Minister that, in his view, it would be wrong to make any assumption about further testing after a CTBT and consequently about the expenditure incurred in autumn 1978 not being wasted if the UK were to have stopped testing before July 1979. Moreover, Dr Owen argued that in so far as an assumption was being made on this score, it seemed to him to be unsound as a basis for deciding to proceed with plans for the proposed

UK test the following July.[89] Such a risk in retrospect looks extremely remote given the state of the treaty negotiations, but as ever officials and Ministers are not blessed with foolproof fortune-telling attributes. As before in 1958, the UK position was that it needed to acquire as much information as it could on advanced nuclear warhead designs *before* a more complete ban on testing came into force.[90] It is also worth noting that the test would also include an ability to produce weapons with a variable yield via a dialling system.[91] UK support for a test ban treaty was thus conditional on securing and confirming sufficient design and associated data first. Despite all this, Defence Secretary Fred Mulley informed the Prime Minister that in seeking approval for funds to prepare for a July 1979 test, he was not seeking to change the UK's position at the CTBT negotiations, or was he making any assumptions about the likelihood that a CTBT would be agreed before July, or of a voluntary moratorium being introduced in anticipations of a CTBT being signed. Nor would he object to UK participation in moratorium if that were deemed the desirable thing to do on wider grounds. The key point for Mulley (as instructed by Macklen) was that unless a decision was made to commit funds now – November 1978 – there would be no prospect of a test being carried out in July.[92] Considerable preparations were required to enable an underground nuclear test to take place as we have already noted, not least of these was drilling the borehole, emplacing the test device and preparing the diagnostic equipment for the test itself. On this basis, the Foreign Secretary agreed that preparations for a test should proceed as proposed – that funds for future tests would be committed in stages as proposed by the Defence Secretary.[93] On the question of the timing and utility of a voluntary testing moratorium, Dr Owen suggested that these matters should be examined in more detail when the timetable for treaty signature and entry into force became clearer.[94] The Prime Minister, having considered the Foreign Secretary's views, concluded that the UK should continue nuclear testing, when necessary, until a CTBT was signed; and therefore, that preparations for the planned July 1979 test should proceed as proposed.[95] Macklen had observed that if this test were successful it would begin to open up possibilities of increasing the yields of these devices, which would certainly open the door to far more exchange with the US on their nuclear devices of a similar nature. The UK test would also include other side experiments that would allow AWRE to gain information of great value in relation to other warhead designs.[96] Quite clearly continued testing was imperative in terms of closer collaboration with the US, contributing to warhead design information, enabling the UK to cope with a CTBT and providing data to help with stockpile maintenance. It would be hard for any Minister to argue against such an ostensibly compelling array of arguments. And so it proved.

The 3 May 1979 saw election of a new Conservative government with Margaret Thatcher as Prime Minister. She was much more suspicious of a test ban treaty and nuclear arms control more generally, and certainly not as enthusiastic about a treaty as Callaghan and Owen had been.[97] When the new Conservative government Ministers met for the first time to consider nuclear policy on

24 May 1979, they concluded inter alia that although the UK should continue to participate in the tripartite test ban treaty negotiations, the UK should, however, ensure, as far as it could, that any tests needed took place before a CTBT entered into force.[98] On 6 July 1979 the new Defence Secretary, Francis Pym, minuted the Prime Minister to inform her that the previous government had authorised preparations for a nuclear test – a test which was now described as 'a key part of our programme of warhead development for deterrent successor systems'. Likely the MOD pitching the message to suit the audience. The device was now in the US, the preparations were well advanced, and the time had come for Ministers to give their final political clearance for the test to proceed.[99]

Pym noted that the possibility of a three-year-long CTBT might impede the UK's warhead development programme had been present for some time. He underlined that Britain was dependent on US goodwill in carrying out these tests – practical preparations took at least nine months to complete. In practice, this meant that the UK had to allow for a period of 12 months between a request for a new test and the desired firing date.[100] If a limited duration CTBT became a reality, then this was not thought likely before late 1980 given the then state of the treaty negotiations in Geneva, which, as we have seen in Chapter 2, were hobbled because of arguments over NSSs. The UK's warhead development programme, aimed at successor systems and modernisation of UK tactical nuclear weapons (the WE177s), would require a least two more successful tests before a test ban came into force. Information acquired from the Chevaline programme had indicated how it ought to be possible to reduce the size and weight of a future warhead without any appreciable loss of yield in order to produce a warhead to defeat future Soviet ABM systems, which would need to be compatible with a high-speed re-entry body rather than in a low-speed re-entry body such as that used for Chevaline.[101] Making allowance for a possible test failure meant that the UK would need to ask the US authorities to seek Presidential clearance for at least three further tests in 1980 – obviously no fretting over how all this might appear publicly unlike their Labour predecessors. As the Defence Secretary advised the Prime Minister in July 1979,

> Our warhead development programme aimed at successor systems and modernisation of our tactical nuclear weapons, would require at least two more successful tests before a (test) ban comes into effect. Making allowance for a possible test failure means that we should ask US authorities to seek Presidential authority for at least three further British nuclear tests in 1980.[102]

The US wanted to submit their own testing programme proposals, along with those from the UK, to the President as soon as possible. Pym therefore sought Thatcher's approval to ask the US to proceed with preparation for two further tests immediately and to leave the third reserve test decision until the UK had the results from the planned August test. Since these tests were deemed by AWRE and MOD to be so important, Pym was prepared to meet the costs from within whatever level of Defence cash provisions that would eventually be made by the Treasury. MOD officials had also recognised that if the UK were to procure a future replacement system for Chevaline from the US, then the Americans would most likely provide full details of the warhead they used in that system. However, it would be unlikely that the UK could produce exact copies of a US design in the timescales required because of the lack of appropriate production methods in the UK. Echoes of the Mark 28 US thermonuclear warhead and UK Red Snow warhead were clearly seen here where anglicisation of the US Mark 28 design presented significant practical engineering problems for the UK in the early 1960s.[103] For this reason, it would be most likely that the UK would need to adopt is own warhead design for a successor system.[104] This helps explain the need for further UK nuclear tests in order to support such a successor system to Polaris/Chevaline.

Lord Carrington, the new Foreign Secretary, agreed with Pym, noting that although a limited CTBT was unlikely to become reality until late 1980, there was a possibility that testing could end earlier than that.[105] Both the US and USSR had an eye on the Second Non-Proliferation Treaty Review Conference in June 1980. It was too soon to say whether the Americans and Russians would stop testing as soon as a treaty had been initialled. However, if the US saw a clear advantage in declaring a moratorium as soon as the treaty had been agreed, the UK would be unwise to assume that President Carter would be dissuaded on the grounds that the UK needed to carry out further tests. Much would depend on the US test programme and whether they had completed their own essential tests. If they had, then the case for a test moratorium would be stronger. If, therefore, there was an essential requirement for two and possibly three tests, the UK should try to show flexibility over timing to ensure that they took place as early as 1980 as technical arrangements allow. However, the MOD view was that a moratorium before a treaty was concluded was not in UK interests – it could end up with the worst of all worlds – an indefinite moratorium and an unratified treaty. AWRE would not be able to accelerate the UK programme since in practical terms it took 40 weeks to mount an underground nuclear test from a go-ahead and that did not include up to eight weeks of further delay while the US government considered the UK request.[106] This period likely encompassed a full technical evaluation of the UK test device, checking especially on the maximum credible yield calculated by AWRE designers, and securing political clearance from the President.

Final political clearance was sought to fire the August test on 27 June 1979. Macklen noted in his submission that if the test were successful, then the UK would be well on the way to providing a British warhead design for a successor system, but CTBT permitting, the UK would need to be able to plan for two or three more tests to ensure that a successor warhead would have sufficient yield, to demonstrate a design for future tactical nuclear warheads, and a spare slot to act as an insurance against a test failure.[107] Only one previous UK nuclear test had failed to fire, and this was the Courser shot of 25 September 1964 when the US-supplied external neutron generators failed to function.[108]

Deep boreholes up to 1,800 feet and about 100 feet in diameter had to be drilled by specialist teams, and the limited number of experienced drilling teams were the bottle-neck in US test preparations at the Nevada Test Site. With a CTBT a possibility, the US was also running an extensive test programme and therefore needed to know UK requirements at an early date in order to accommodate the British in its schedule. In fact, if the UK wanted to test in mid-1980, the US sought British approval by end of July 1979 for the drilling preparations to start. Macklen observed that if a three-year CTBT became reality by the end of 1980, he would recommend that the UK should immediately ask the US to make preparations for two British tests in the mid-1980 and commit the necessary money to pay for the US preparations. He added that preparations for a third test could be delayed until early autumn, by which time the result of the August 1979 test would be available. If this were a failure, then a third test would become a necessity. However, the dates which might be offered by the US would be likely adversely affected because of a shortage of US funds until the start of the Fiscal Year 1980. The administrative problem here was that the US Department of Energy had too little money to retain a drilling crew at Nevada until 1 October and it was going to be paid off in September. Consequently, time would be lost after 1 October in recruiting the new crew needed to prepare for the UK tests. The DOE asked MOD informally whether it could consider advance funding on account, but revert later to billing in arrears as was the more normal practice. Since the UK was under political pressure, particularly in view of the CTBT situation, to complete its test programme as early as possible in 1980, the MOD was keen to respond favourably to the US request.[109] Approval was given to make the payment. Ministerial authority was also provided in July 1979 for three further tests in 1980.[110] As usual Presidential authority was required as the US DOE would be unable to begin preparations until this had been received. MOD officials hoped that in view of the CTBT negotiations any delay could be minimised; and were even prepared to commit funds to enable the preparations to start early in the event of a serious delay in obtaining President Carter's approval for the three tests.

In the end, the Nessel test was fired on 29 August 1979 – the immediate assessment of the senior AWRE scientist present at the site was that the test had been successful, which was later confirmed by the preliminary numerical data. The yield of the shot, which was recorded seismically, appeared to be in accordance with the prediction and the device's main components all appeared to have operated very well. Although it was still early days with much data (including radioactive samples created by the explosion recovered from the cavity by drill back) still to be analysed at AWRE, the outlook was promising and gave confidence both in AWRE's design ability and in the prospects for developing a warhead to meet a future requirement for a successor deterrent system.[111]

Information in The National Archives on the five UK nuclear tests conducted between 1980 and 1982 is very sparse. Many of the relevant files are either closed permanently or redacted. There is little on the tests' objectives and outcomes. In general, we can say that these tests were linked both to strategic and tactical nuclear weapons as well as broader scientific enquiry into nuclear explosion physics. In January 1980 Foreign Secretary Lord Carrington questioned why the MOD thought it advisable to develop two options for the warhead for Trident.[112] He also noted that a test ban treaty might impact British testing plans in 1981, a point acknowledged by the Defence Secretary. There is no detail in the archives (at least found by the author) explaining why the MOD wanted to explore two warhead designs for Trident. One could speculate that, given previous experience with Chevaline, the choice might have been about different weights, which would have an impact on range; or was between a more advanced design that possibly used less fissile material and whose prospects might have been on the edge of achievability and a more conservative design that was much more likely to be successful and perhaps easier to maintain if tests were to be prohibited. It also seems that MOD was thinking in terms of a higher yield for its warhead for Trident than the Americans were planning for theirs.[113] Again, we do not know why this requirement existed. Carrington also noted, significantly, that the last time Ministers had considered the UK testing programme, the requirement was for three tests in 1980. This had been on the basis that at least two more tests were needed after the one in August 1979. As he understood it, this meant that there was no question of a CTBT preventing the UK from having its own warhead for a Polaris successor. This implies that AWRE would have all the design information validated by tests to produce a new warhead for Trident. Indeed the independent Nuclear Advisory Panel, Chaired by Lord Penney, had firmly concluded that nuclear testing would be necessary to justify the introduction of a new warhead to the UK's stockpile.[114]

The first of these three tests in 1980 – Colwick (UK device code name was Dingbat) was fired on 26 April. Dingbat appears to have been an 'experimental device which might almost match the current US performance'.[115] Preliminary indications immediately after the test suggested that the Dingbat device achieved its designed nuclear yield.[116] All we can determine about Dutchess, the next UK test on 24 October 1980, was that it sought to test a mechanism for varying the yield of future tactical nuclear weapons.[117] The third and final of the three tests planned for 1980 took place on 17 December. All we can say about the purposes of this device, Hurdle Prime,[118] was that it sought to evaluate a possible warhead for the system to replace Polaris.[119] By late spring 1980, both the MOD and FCO were satisfied that there was no prospect of a CTBT in a timescale short enough to affect the UK's ability to test the new warhead for Trident adequately when it was ready.[120] Although by November the MOD were now insisting that a resumption of testing would be essential after 1986.[121] We have slightly more detail, but not much, on the remaining two UK nuclear tests conducted in the period covered by this book: Rousanne (UK codename Dingbat II) 12 November 1981 and Gibne (UK Codename Hurdle Prime) on 25 April 1982. Dingbat II included a yield-switching mechanism, which was shown to be satisfactory, and would be of value in a new tactical warhead to replace the WE177s. AWRE also obtained some new information to widen its technology base. The test was important because the device incorporated some new warhead features that were beyond the UK's then-present powers to evaluate quantitatively.[122] All the objectives of the test were successfully met.[123] Hurdle Prime was intended to inform an important part of AWRE's progress towards designing warheads for the new Trident missiles in the 1990s.[124] Preliminary indications were that the operation went according to plan and that the test was successful.[125] So overall the UK was well placed as a result of all of these tests to move towards a design for a production warhead for Trident.

Conclusion

Despite some inevitable anxieties that a test ban treaty or testing moratorium might constrain or prevent the UK's ability to conduct essential nuclear tests at Nevada, it seems that the threat was never that serious given the continuing challenges facing the treaty negotiations in Geneva. The actual prospects of a ban treaty emerging from the tripartite negotiations look, in retrospect, to have been remote. That said, it is in the very nature of British officials and the civil service to worry about contingencies, and therefore to plan and act accordingly. Even at the time, Ministers and officials must surely have realised that a test ban was unlikely, especially once the row over the Soviet insistence that the UK would need to host ten seismic monitoring stations on UK territory as part of the treaty's Separate Verification Agreement, including dependent territories, proved a major obstacle to progress in 1979 and 1980. This issue was discussed in greater detail in Chapter 2. Nevertheless, the nuclear tests that the UK conducted between 1978 and 1982 were clearly essential for the modernisation of the UK's strategic and tactical nuclear weapons systems as well as generating a considerable amount of scientific data on nuclear explosion physics that would have helped the UK to maintain its nuclear stockpile under a test ban regime. It seems clear, therefore, that the MOD's key interests in the CTBT negotiations as outlined in July 1977 were preserved: firstly, the maintenance of the technical effectiveness of existing UK nuclear forces and hence the credibility of deterrence; and, secondly, the maintenance of options for improving existing UK nuclear forces and for successor systems.[126]

Notes

1   TNA DEFE 13/1768, V.H.B. Macklen, Nuclear Warhead Research and Capability of AWRE, 6 June 1976. See also TNA DEFE 13/2826/2, R.J. Moon, Security Policy Department, FCO to N.K. Witney, D Nuc Pol/Sy, Visit to Aldermaston, 25 April 1991.
2   TNA PREM 19/14, The Nuclear Warhead Test Programme, MOD, 17 May 1979.
3   TNA CAB 130/1158, MISC 1 (81) 3, Cabinet Official Group on International Aspects of Nuclear Defence Policy, Nuclear Test Ban Policy: Report, Note by the Secretary, Annex D Technical Aspects of a Low Threshold Test Ban, 6 February 1981.
4   TNA DEFE 19/272, V.H.B. Macklen, DCA (PN) to PUS, Maintenance of the Nuclear Capability in event of Option B, Maintenance of British Nuclear Weapon Capability at AWRE if the UK buys US Poseidon/Option B, 15 June 1973.
5   Matthew Jones, *The Official History of the UK Strategic Nuclear Deterrent, Volume II: The Labour Government and the Polaris Programme, 1964–1970*, London, Routledge, 2017, p. 302. The JIC established deterrence criteria in 1962, but Jones notes that the criteria were never updated, or much referred to in internal debates.

6 TNA DEFE 13/1039, Chief Scientific Advisor to Secretary of State, Future Nuclear Warhead Research Programme, 17 May 1976.

7 TNA DEFE 13/1039, APS/CSA to PS/SoS, CSA/208/76 of 17 May 1976 and 18 May 1976.

8 TNA DEFE 13/1039, Roy Mason to CSA, Future Nuclear Warhead Research Programme, 19 May 1976.

9 TNA DEFE 13/1768, Chief Scientific Advisor to Secretary of State, Nuclear Weapon Research, 6 September 1976.

10 TNA DEFE 13/1768, Nuclear Warhead Research and the capabilities of AWRE, Report by DCA (PN), Annex A to CSA 389/76, 6 September 1976.

11 TNA DEFE 13/1768, V.H.B. Macklen, DCA (PN) to Secretary of State, Authority to Proceed with Preparations for an Underground Nuclear Test in Autumn 1978, 24 February 1977. It is worth noting the absolutely pivotal role that Macklen played in British nuclear weapons policy from the 1950s through to the 1970s, with a brief time away at the UKAEA's Reactor Group in the mid-1960s. Mac, as he was known, always made a clear case for further work on nuclear weapons and his opinions carried much weight inside the MOD and elsewhere in Whitehall and in the US. Macklen was instrumental in all UK nuclear weapons decision-making in this period. He did not suffer fools gladly and did not get on with Sir Solly Zuckerman when the latter was first Chief Scientific Advisor at the MOD and then later when he served as Chief Scientific Advisor to the Government.

12 TNA DEFE 13/1768, R.L.L. Facer, PS/Secretary of State, Nuclear Weapon Research, 7 October 1976.

13 Kristan Stoddard, *The Sword and the Shield: Britain, America, NATO and Nuclear Weapons, 1970–1976*, London, Palgrave Macmillan, 2014, p. 162; TNA DEFE 19/207, The Project Definition Study of Super Antelope (KH 793) Report of the KH 793 Project Review Board, 3 October 1972.

14 National Security Archive, George Washington University, US Department of State Information Memorandum, PM – Ronald I. Spiers to Deputy Secretary, US-UK Nuclear Co-operation, Letter from Deputy Secretary of Defense Rush, 26 July 1972.

15 TNA DEFE 19/273, KH 793 Project Status, November 1975.

16 TNA DEFE 19/273, KH 793 Project Status, November 1975.

17 TNA DEFE 13/1039, V.H.B. Macklen, DCA (PN) to PS/SoS, Chevaline Nuclear Warhead Tests, 27 February 1976.

18 TNA PREM 16/1181, Roy Mason to Harold Wilson, Polaris Improvement Programme – Nuclear Testing, 28 May 1975.

19 The number of ABM warheads required to destroy in-coming warheads.

20 TNA DEFE 13/1039, Secretary of State for Defence to Prime Minister, Polaris Improvement Programme – Nuclear Testing, 4 March 1976.

21 TNA DEFE 13/1039, Patrick Wright, 10 Downing Street to J.F. Mayne, MOD, Polaris Improvement Programme – Nuclear Testing, 8 March 1974.

22 TNA DEFE 13/1768, A.P. Hockaday, DUS (P) to PS/Secretary of State, Chevaline and Nuclear Testing, 19 July 1976.

23 TNA DEFE 13/1768, J.F. Mayne, PS/Secretary of State to DCA (PN), Polaris Improvement Programme – Nuclear Testing, 23 July 1976.

24 TNA DEFE 13/1768, V.H.B. Macklen, DCA (PN) to PS/SoS, Polaris Improvement Programme – Nuclear Testing, 26 July 1976.

25 TNA DEFE 13/1768, V.H.B. Macklen, DCA (PN) to PS/Secretary of State, Chevaline Nuclear Test, 27 July 1976.

26 TNA DEFE 13/1768, V.H.B. Macklen, DCA (PN) to PS/Secretary of State, Nuclear Tests, 29 July 1976.

27 TNA DEFE 13/1768, Meeting of Ministers, The Nuclear Deterrent, Minutes of a Meeting Held at 10 Downing Street, Thursday 29 July 1976.

28 TNA DEFE 13/1768, A.P. Hockaday, DUS (P) to PS/Secretary of State, Forthcoming Nuclear Test, 3 August 1976.

29 TNA FCO 66/875, CTB Negotiations: Implications for Defence Interests, Note by the Ministry of Defence, 1 July 1977; TNA DEFE 23/1344, Report of a Meeting Held in the US State Department on the Afternoon of Monday, 14 March 1977 to Discuss Comprehensive Test Bans.

30 TNA DEFE 13/1768, V.H.B. Macklen, DCA (PN) to Secretary of State, Authority to Proceed with Preparations for an Underground Nuclear Test in Autumn 1978, 24 February 1977.

31 TNA DEFE 13/1768, V.H.B. Macklen, DCA (PN) to Secretary of State, Authority to Proceed with Preparations for an Underground Nuclear Test in Autumn 1978, 24 February 1977.

32 TNA DEFE 13/1768, Fred Mulley to Prime Minister, Requirement for a Further Underground Nuclear Test, 25 February 1977.

33 TNA DEFE 13/1768, Patrick Wright, 10 Downing Street to Roger Facer, MOD, Requirement for a Further Underground Nuclear Test, 9 March 1977.

34 TNA DEFE 13/1768, V.H.B. Macklen, DCA (PN) to PS/SoS, CTBT: Meeting with Foreign Secretary, 1 April 1977.

35 TNA FCO 66/897, D.C. Fakley, Director, D Sc 6, MOD to J.C. Edmonds, ACDD, FCO, UK Programme for Maintaining the Effectiveness of the Deterrent, 20 April 1977.

36 TNA DEFE 13/1768, V.H.B. Macklen, DCA (PN) to Secretary of State, The Next British Nuclear Test, 26 May 1977; TNA DEFE 13/1768, Fred Mulley to Prime Minister, Planned Date for the Possible Next British Nuclear Test, 2 June 1977.

37 TNA DEFE 13/1768, Patrick Wright, 10 Downing Street to Roger Facer, MOD, Planned Date for the Possible Next British Nuclear Test, 7 June 1977.

38 TNA PREM 16/1183, Patrick Wright, Private Secretary No 10 Downing Street to Sir Clive Rose, Cabinet Office, Test Ban Negotiations, 7 June 1977.

39 TNA DEFE 13/1769, Implications for Arms Control Annex B Paragraph 3 a, Undated.

40 TNA DEFE 23/219, V.H.B. Macklen, DCA (PN) to PS/PUS, British Nuclear Test Programme, Attachment paragraph 2, 8 November 1977.

41 TNA DEFE 13/1769, V.H.B. Macklen, DCA (PN) to Secretary of State, British Nuclear Test – March 1978, 18 July 1977.

42 TNA 66/1081, R.L.L. Facer, MOD to K.R. Stone, 10 Downing Street, Polaris Improvement Programme – Nuclear Testing, 9 February 1978.

43 Kristan Stoddard, *The Sword and the Shield: Britain, America, NATO and Nuclear Weapons, 1970–1976*, London, Palgrave Macmillan, 2014, p. 164.

44 TNA DEFE 13/1478, British Nuclear Test Programme, V.H.B. Macklen, 19 April 1978.

45 TNA DEFE 13/1477, British Nuclear Test Planned for March 1978, V.H.B. Macklen, 17 November 1977. The Fondutta test was postponed from March to April.

46 TNA DEFE 23/222, Ministry of Defence Chevaline Steering Committee, Minutes of Meeting Held on Friday 15 December 1978, 15 December 1978.

47 TNA DEFE 13-1478, British Nuclear Test Programme, 27 April 1978 and TNA DEFE 13/1478 e24; Worded as "*a Significant Advance in Design*" in Macklen's Draft, 19 April 1978.

48 TNA DEFE 24/1361, AWRE, Aldermaston Classification Notice No. 43, Codenames of Underground Tests of UK Nuclear Devices at NTS, S.K. Hutchison, 7 March 1980.

49 TNA FCO 46/1830, F. Mulley to Prime Minister, British Nuclear Test – 11 April 1978, 27 April 1978.

50 TNA DEFE 25/433, Frank Cooper to Secretary of State, The Defence Nuclear Programme, 17 May 1979.

51 TNA DEFE 23/219, Chevaline Progress Report, July 1978.

52 TNA FCO 46/1830, F. Mulley to Prime Minister, British Nuclear Test Programme, 27 April 1978.

53 TNA FCO 46/1830, W.J.A. Wilberforce, Defence Department to Mr Moberly and Private Secretary, British Nuclear Test Programme, 3 May 1978.

54 TNA FCO 46/1830, F. Mulley to Prime Minister, British Nuclear Test Programme, 27 April 1978.

55 TNA FCO 46/1830, David Owen to Prime Minister, British Nuclear Test Programme, 9 May 1978.

56 TNA FCO 46/1830, B.G. Cartledge, 10 Downing Street to R.L.L. Facer, MOD, British Nuclear Test Programme, 8 May 1978.

57 TNA FCO 46/1830, R.L.L. Facer, MOD to B.G. Cartledge, 10 Downing Street, British Nuclear Test Programme 27 May 1978; TNA FCO 66/1081, Fred Mulley to Prime Minister, British Nuclear Test Programme, 9 June 1978.

58 TNA DEFE 19/181, Background Note, M. Nutting, 21 November 1978.

59 TNA DEFE 19/181, V.H.B. Macklen, DC(PN) to Secretary of State, British Nuclear Test, 1978.

60 TNA DEFE 13/1478, British Nuclear Test Programme, V.H.B. Macklen, 19 April 1978.

61 TNA DEFE 13/1477, British Nuclear Test Programme, V.H.B. Macklen, 21 November 1977.

62 TNA DEFE 13/1478, British Nuclear Test Programme, 27 April 1978.

63 TNA DEFE 13-1478, British Nuclear Test Programme, V.H.B. Macklen, 19 April 1978. TNA DEFE 24/1361, Background Note, M. Nutting, 21 November 1978.

64 TNA DEFE 25/335, British Nuclear Test, V.H.B. Macklen, 20 November 1978.

65 TNA DEFE 24/1361, D.C. Fakley, D/DSc 6 to Head of DS2, 1980 Underground Nuclear Test Programme, 1 August 1979.

66 The test was eventually conducted on 29 August – codenamed Nessel.

67 TNA DEFE 24/1361, Background Note, M. Nutting, 21 November 1978.

68 TNA DEFE 19/181, British Nuclear Test Programme, V.H.B. Macklen DCA (PN) to PS SoS, 6 June 1978.

69 TNA DEFE 19/181, V.H.B. Macklen DCA (PN) to PUS, British Nuclear Test Programme Note for the Record, 15 August 1978.

70 TNA DEFE 19/181, Foreign Secretary to the PM, British Nuclear Tests Programme, 31 October 1978.

71 TNA DEFE 19/181, V.H.B. Macklen, DCA (PN) to PS/Secretary of State, 2 November 1978.

72 TNA FCO 66/1081, P.H. Moberly, British Nuclear Test Programme. Mr Wilberforce's Submission of 10 November, 14 November 1978.

73 TNA DEFE 19/181, Secretary of State for Defence to Prime Minister, British Nuclear Test Programme, 3 November 1978.

74 TNA DEFE 13/1477, The UK Test on 16 March 1978, 30 November 1977.

75 TNA DEFE 19/181, Fred Mulley to Prime Minister, British Nuclear Test Programme, 3 November 1978.

76 TNA DEFE 19/181, V.H.B. Macklen, DCA (PN) to Secretary of State, British Nuclear Test Programme, 22 November 1978.

77 TNA DEFE 19/181, V.H.B. Macklen DCA (PN) to Secretary of State, British Nuclear Test Programme, 22 November 1978.

78 TNA DEFE 19/181, V.H.B. Macklen to PUS, British Nuclear Test Programme Note for the Record, 15 August 1978.

79 TNA DEFE 19/181, V.H.B. Macklen to PS/SoS, British Nuclear Test Programme, 23 October 1978.

80 TNA DEFE 13/1478 and see also in TNA DEFE 25/335, British Nuclear Test Programme, 3 November 1978.

81 TNA DEFE 13/1478, British Nuclear Test Programme, 3 November 1978.

82 All UK test designs for Nevada had to be shared with the US before formal Presidential authority would be given to proceed, and this included an assessment of maximum credible yield. The US nuclear weapons labs undertook an independent assessment of the UK design, which was a necessary preliminary to its acceptance in principle by the range authorities at Nevada as a device suitable for testing. TNA DEFE 13/1039, F.H. Panton, ACSA (N) to AUS (OR), Further Nuclear Test, 3 October 1974.

83 TNA DEFE 13/1478, British Nuclear Test Programme, V.H.B. Macklen, 25 October 1978.

84 TNA DEFE 25/335, British Nuclear Test, V.H.B. Macklen, 20 November 1978.

85 Hansard, Volume 988, Column 1235, Tuesday 15 July 1978.

86 TNA DEFE 19/181, Ken Johnston, AD/DSc6 to D. Bryars, AUS (D Staff), Grey Area Systems, 9 November 1978.

87 TNA FCO 46/1830, C.L.G. Mallaby to Mr Moberly and Private Secretary, British Nuclear Test Ban, 27 October 1978.

88 TNA DEFE 19/181, David Owen to Prime Minister, British Nuclear Test Programme, 31 October 1978.

89 TNA FCO 66/1081, David Owen to Prime Minister, British Nuclear Test Programme, 16 November 1978.

90 TNA DEFE 19/181, V.H.B. Macklen DCA (PN) to Secretary of State, British Nuclear Test Programme, 22 November 1978.

91 TNA FCO 46/1830, C.L.G. Mallaby, Arms Control and Disarmament Department to Mr Moberly and Private Secretary, British Nuclear Test Programme, 27 October 1978.

92 TNA DEFE 19/181, Fred Mulley to Prime Minister, British Nuclear Test Programme, 23 November 1978.

93 TNA FCO 66/1081, Fred Mulley to Prime Minister, British Nuclear Test Programme, 23 November 1978 and G.G.H. Walden to Defence Department, British Nuclear Test Programme, 27 November 1978.

94 TNA DEFE 19/181, G.G.H. Walden to B.G. Cartledge, 10 Downing Street, British Nuclear Test Programme, 28 November 1978.

95 TNA FCO 66/1081, Bryan Cartledge, Private Secretary No.10 to R.L.L. Facer, Foreign and Commonwealth Office, UK Nuclear Test Programme, 1 December 1978.

96 TNA DEFE 19/181, V.H.B. Macklen, DAC (PN) to Secretary of State, British Nuclear Test – 20 November 1978, 23 November 1978.

97 TNA PREM 19/212, John Hunt, Cabinet Secretary to Prime Minister, Comprehensive Test Ban, Comprehensive Test Ban Negotiations, May 1979; B.G. Cartledge, 10 Downing Street to M.J. Vile, Cabinet Office, Comprehensive Test Ban, 9 May 1979. The Private Secretary noted in his letter that the PM was 'concerned' whereas Thatcher's own manuscript comment on the Cabinet Secretary's brief is 'very unhappy and suspicious'. And the word 'very' is underlined.

98 TNA CAB 130/1109, MISC 7 (79) 1st Meeting Cabinet Nuclear Defence Policy, Note of a Meeting Held at 10 Downing Street on 24 May 1979.

99 TNA DEFE 19/181, Defence Secretary to Prime Minister, British Nuclear Test Programme, 6 July 1979.

100 TNA PREM 19/14, R.L.L. Facer, MOD to B.G. Cartledge, 10 Downing Street, Nuclear Warhead Test Programme, 24 May 1979.

101 TNA PREM 19/14, R.L.L. Facer, MOD to B.G. Cartledge, 10 Downing Street, Nuclear Warhead Test Programme, 24 May 1979.

102 TNA DEFE 24/1361, Defence Secretary to Prime Minister, British Nuclear Test Programme, 6 July 1979.

103 John R. Walker, *British Nuclear Weapons and the Test Ban 1954–1973: Britain, the United States, Weapons Policies and Nuclear Testing: Tension and Contradictions*, Farnham, Ashgate, 2010, p. 98.

104 TNA PREM 19/14, R.L.L. Facer, MOD to B.G. Cartledge, 10 Downing Street, Nuclear Warhead Test Programme, 24 May 1979.

105 TNA DEFE 19/181, Carrington to Prime Minister, British Nuclear Test Programme, 12 July 1979.

106 TNA DEFE 19/181, V.H.B. Macklen, DCA(PN) to PUS, British Nuclear Test Programme, 23 July 1979.

107 TNA DEFE 19/181, V.H.B. Macklen, DCA (PN) to Secretary of State, British Nuclear Test Programme, 27 June 1979.

108 TNA DEFE 24/291, Cabinet Nuclear Requirements for Defence Committee, British Programme of Underground Nuclear Tests, Note by UKAEA, 1965.

109 TNA DEFE 19/181, D.C. Fakley, D/Dsc 6 to Head of DS 2, 1980 Underground Nuclear Test Programme, 1 August 1979.
110 TNA DEFE 24/1316, V.H.B. Macklen, (DCA) PN to Duane Sewell, Assistant Secretary for Defence Programmes, US Department of Energy, 24 July 1979.
111 TNA DEFE 24/1316, CSA to Secretary of State, British Underground Nuclear Test, 30 August 1979.
112 TNA PREM 19/159, Note to the Prime Minister, British Nuclear Test Programme, 9 January 1980.
113 TNA PREM 19/159, Carrington to Prime Minister, British Nuclear Test Programme, 3 January 1980.
114 TNA PREM 19/693, Robert Armstrong to Prime Minister, Nuclear Advisory Panel, 20 May 1980.
115 TNA PREM 19/14, R.L.L. Facer, MOD to B.G. Cartledge, 10 Downing Street, Nuclear Warhead Test Programme, 24 May 1979.
116 TNA PREM 19/14, R. Armstrong to Prime Minister, British Underground Nuclear Test, 28 April 1980.
117 TNA CAB 128/68/13, Cabinet, CC (80), Conclusions of a Meeting of the Cabinet, p. 9, 23 October 1980.
118 TNA PREM 19/417, PPS 10 Downing Street to B.M. Norbury, MOD, British Nuclear Test Programme, 12 February 1981.
119 TNA PREM 19/159, Note to Prime Minister, British Nuclear Test Programme, 9 January 1980.
120 TNA PREM 19/159, R.L. Wade-Grey to Prime Minister, MISC 7 The Future of the United Kingdom Strategic Deterrent: The Present Position, 30 May 1980.
121 TNA FCO 66/1473, A. Reeve, Arms Control and Disarmament Department to Mr P.H. Moberly, MISC 1 CTB, 20 November at 3.00 pm, 19 November 1980.
122 TNA PREM 19/694, Secretary of State for Defence John Nott to Prime Minister, British Underground Nuclear Test, 11 December 1981.
123 TNA PREM 19/694, Clive Whitmore, 10 Downing Street to David Omand, MOD, British Nuclear Test, 17 December 1981.
124 TNA PREM 19/417, Robert Armstrong to Prime Minister, British Nuclear Tests, 9 September 1980.
125 TNA PREM 19/695, D.T. Piper, MOD to A.J. Coles, 10 Downing Street, 29 April 1982.
126 TNA FCO 66/875, CTB Negotiations: Implications for Defence Interests, Note by the Ministry of Defence, 1 July 1977.

4 Stockpile Reliability and Safety and the Test Ban

UK concerns

Introduction

Until the 1970s, the reliability of the UK nuclear weapons stockpile under a CTBT never seems to have been an issue during the early days on test ban treaty negotiations in the 1950s.[1] Whether this was attributable to the lack of technical knowledge on the ageing and interactions of fissile material and other non-nuclear components in a weapon, or the relatively short design lives of the weapons designed and produced in the late 1950s and early 1960s is unclear. We can be sure, however, that as continued underground testing was permissible under the Partial Test Ban Treaty (PTBT) agreed in 1963, the issue never arose since had a new warhead been required in the UK, or a major reliability issue had arisen, then the option to test was available. That said, Harold Wilson's Labour Government had declared publicly that there would be no strategic successor to Polaris and even though the UKAEA wished to conduct research tests, the UK consequently refrained from testing from 1965 until 1974. The lessons from this experience were to be of interest to the US as discussions about weapon reliability under a test ban assumed greater salience in the tripartite test ban treaty negotiations. Two key questions arise. First, if warhead reliability was not an issue in 1958, why was this so and why did it emerge later as a serious problem in the late 1970s – genuinely better technical understandings of stockpile reliability and safety issues acquired over 20 years, or innate US opposition to a test ban, or a bit of both? Second, what was the UK reaction? Furthermore, how did the UK intend to deal with this problem given its dependencies on the US created by the 1958 Mutual Defence Agreement (MDA) and how would this issue shape British attitudes to a CTBT? One key question facing Ministers and officials in the late 1970s and into the 1980s was whether a CTBT was even desirable at all under conditions where weapon reliability was uncertain and whether the compromises required to secure this fatally undermined a test ban treaty's non-proliferation benefits. We also need to keep in mind that by this stage in its weapons programme British nuclear weapons were produced to very close engineering tolerances with a wide range of accurately specified materials, some of which were very reactive chemically or radioactive and very precise geometries. To economise both in warhead size and weight warhead

DOI: 10.4324/9781003375708-4

designs were marginal in the sense that a minor degradation in the performance of a warhead could cause the warhead to fail completely.[2] This consideration sat very much in the front seat during debates over the competing and at times diametrically opposed considerations over the merits and demerits of a nuclear test ban treaty. Britain's basic nuclear warhead technology was its own, but it was heavily reinforced by information exchanges with the US laboratories at Los Alamos and Lawrence Livermore. These exchanges materially reduced the cost of the technological base of the British programme, essential if the UK were to remain a nuclear weapons state. For this reason, the continuation of the 1958 Mutual Defence Agreement and the 1963 Polaris Sales Agreement was thus of vital importance to the UK; and it was therefore essential that nothing was done in the context of CTBT negotiations that would threaten them.[3] As Ministers and officials recognised, the UK was heavily reliant on the US for its nuclear warhead technology and as such it would be inconceivable that the UK on its own could come up with a solution to the problem of stockpile safety and reliability that was different from the one adopted by the US.[4] In addition, as the UK's stockpile was considerably smaller and relied on only two basic designs, an underlying British fear was that a CTBT would erode UK nuclear capabilities much more rapidly than that of either the US or USSR.[5] Previous choices – such as the reestablishment of the close nuclear relationship with the US in 1958, the decision that there would be no strategic successor to Polaris, the need to embark on the Polaris Improvement Programme in light of perceived threats posed by Soviet anti-ballistic missile (ABM) developments and the self-imposed testing moratorium shaped the context for the UK's nuclear weapons and arms control policies in the 1970s.

The source of the problem

Following the Cabinet Office Official Group on International Aspects of Nuclear Defence Policy's (known for short as GEN 63) examination of the nuclear arms control policy of the new Carter Administration in April 1977, the UK decided to seek participation at the outset with the US on test ban treaty negotiations with the Soviet Union.[6] This group, consisting of Cabinet Office, MOD and FCO senior officials, had reviewed the problem of stockpile safety and reliability in July 1977 before the negotiations began. In reviewing the pros and cons of a CTBT for the UK, Ministers (basically the Prime Minister, Defence and Foreign Secretaries) were advised that a treaty could be harmful to UK national objectives – maintenance of the Western deterrent. It was important that nuclear arms control did not alter the strategic balance of conventional and theatre nuclear forces in Europe; and, moreover, that arms control would not hamper the maintenance of the technical effectiveness of the UK deterrent. Options to maintain or improve current UK nuclear forces and potential successor systems to Chevaline/Polaris would need to remain open.[7] Britain harboured concerns too about Soviet cheating: even with the then-envisaged verification system for a test ban treaty, the USSR could conduct small-yield tests of significant military value without being detected. If they chose to play it safe by limiting yields to one kiloton, then Soviet weapons laboratories could do much useful work on developing fission weapons and weapons for tactical and anti-ballistic missile purposes. If they were prepared to take greater risks by testing up to five kilotons, they could continue to develop thermonuclear warheads to the point where only one full-scale confirmatory test would be required before putting them into service.[8]

MOD officials subsequently prepared a new paper for GEN 63. This noted that the ability to maintain existing stocks of British nuclear weapons in a safe and serviceable condition depended upon the competence of the highly skilled scientists and technologists who provided the stockpile surveillance service. As long as the scientific teams were kept together, these skills would be available, but under the conditions of a CTBT, the UK would no longer be able to demonstrate that this expertise remained effective.[9] MOD noted that in cases where major remedial measures were needed to rectify faults in an in-service weapon, it would not be possible to prove the effectiveness of the changed design by a confirmatory underground test. Such tests were considered essential since major changes could not always be validated on the basis of theoretical assessment and laboratory experimentation with sufficient confidence to satisfy Royal Navy or RAF requirements. Until this point the UK had never had the need for a confirmatory test in such circumstances, so what we are seeing here is a projection of what could rather than what would happen at some future point in time. If a new warhead design was required for either a strategic system or tactical weapons, then this could be severely affected if a CTBT was to enter into force quickly. From the results obtained from the Fallon and Banon tests carried out in 1974 and 1976, respectively, an improved warhead was developed for Polaris. AWRE believed that it would be possible to make this warhead more effective operationally by introducing modifications, which would have to be approved by a further test that was originally planned for October 1978, which as we saw in Chapter 3 was brought forward to March 1978 through concerns about an early test ban treaty coming into force. Furthermore, a test ban would also limit the UK's ability to advance its warhead technology because of the adverse effects on the retention of skilled personnel at AWRE. But as we shall see in Chapter 5, it would be the retention of different types of expertise that would in practice pose more of a threat to the weapons programme. This disadvantage would, however, be reduced if the US were willing to share up-to-date information of their warheads tested since 1963. However, even then AWRE still thought that it would be essential to reassure itself that apparently minor changes in manufacturing processes and material specifications would not alter the effectiveness of a radically new warhead built to American designs. In such circumstances, AWRE would still want to conduct a confirmatory test. At that stage – mid-1977 – the UK thought that the US position on stockpile reliability seemed to be based on the assumption that they would be able to maintain their stocks satisfactorily. That would not be enough for the UK: it would need to assure itself that the US assumption was well-founded and that the UK would be able to do the same for its modest stockpile of about 460 weapons.[10] Underlying all this was the fear that the UK and Western nuclear deterrents' credibility would wane over the years of a CTBT, gradually shifting the strategic balance in favour of the USSR, which would be better able to retain its experts in the weapons laboratories given the closed and command nature of Soviet society. And this would be made even worse if peaceful nuclear explosions (PNEs) were allowed to continue since this would allow Soviet weapons designers to retain their expertise under the guise of a 'peaceful' programme. Worryingly, the US Energy Research and Development Administration, the predecessor organisation to the Department of Energy, which took over reasonability for the US weapons labs in October 1977 and the Department of Defense, were already indicating in late May 1977 that they were concerned that a long-term ban on testing, such as ten years or more, would seriously affect both the reliability of the US nuclear stockpile and their ability to keep their weapons design teams and knowledge intact.[11] This anxiety level would increase further in the coming year as we shall see, with ominous consequences for the treaty.

MOD's note on the implications for UK defence requirements was duly discussed by GEN 63 at its tenth meeting on 28 June 1977.[12] The note did not go down too well with other committee members. We see that the meeting notes record that the paper lacked balance as it contained none of the arguments in favour of a CTBT – it was also very polemical in style. On the other hand, it was clear to GEN 63 that it had to be recognised that the arguments in the MOD paper were of considerable importance, especially the problem of loss of expert staffs if testing were no longer an option. It was probable that the US would lose its expertise first, followed by the UK and finally the USSR. For this reason, a treaty that allowed withdrawal after five years was likely to be the worst option of all from a Western point of view since the USSR would be able to keep its design teams together longer and the US could not, nor could it respond quickly to a renewed phase of testing. Concerns over a five-year treaty are odd in view of what was to happen on the question of duration the following year, but we are getting ahead of ourselves. A subsequent GEN 63 meeting concluded that a revised paper on the advantages and disadvantages of a test ban treaty need not be submitted to Ministers before the opening of the tripartite negotiations on 13 July. Many of its arguments had already been included in the group's initial report to Ministers in GEN 63 (77)1 at the end of April and Ministers' decision to participate in CTBT consultations would not be affected by a further paper in July.[13] A further revised version of the advantages/disadvantages of a CTBT paper was nevertheless finalised on 15 July; it also included an MOD assessment of the implications of a CTBT for UK defence interests.[14] The core arguments remained unchanged, and they would essentially remain so throughout subsequent policy discussions during the treaty negotiations. Officials, therefore, had to find ways of squaring the circle. This was not going to be easy as we shall see because the basic objectives were diametrically opposed when stated starkly: nuclear defence or nuclear arms control. Squaring circles was not likely going to be possible, so the prospects for a CTBT were thus poor before the formal negotiations even began.

Come September 1977 the Prime Minister requested that GEN 63's paper should be discussed with the US before the next round of tripartite negotiations, which were scheduled to start on 3 October 1977.[15] GEN 63 decided to prepare a sanitised version to pass to the State Department and the Department of Defense and that the US should be told that it was a revised version of the earlier paper shared with Washington, but not of the Prime Minister's comments.

In talks with the US Department of Energy (DOE) in autumn 1977, the UK hoped to resolve divergent views on stockpile reliability, but DOE officials reported that the US was still not ready to discuss this issue yet with UK experts. For this reason, it had not proved possible in the bilateral talks with the Americans on 16 September to discuss the State Department's assessment on the effect of a test ban on the ability to maintain stocks of existing nuclear weapons in a safe and serviceable condition. GEN 63 felt that it was important to be sure that the US technical agencies subscribed to the optimistic conclusions of the upbeat assessment, but as the US paper was said to be an inter-agency one, UK enquiries in Washington would have to be pursued discreetly.[16] Officials agreed on the need to consider this issue further with the appropriate US authorities and GEN 63 chairman Sir Clive Rose (a senior FCO official on secondment to the Cabinet Office) would pursue this matter with the MOD. There is no further reference to this matter in GEN 63's remaining meetings in 1977. London then assumed thereafter that the US had a sound database on which to assess stockpile safety and reliability. However, FCO and MOD officials also noted at the time that the US position had first appeared back in the spring,

> to be based on the assumption that they will be able to maintain their weapon stocks satisfactorily, but we need to satisfy ourselves that this assumption is technically well founded and that we shall be able to do the same for our own stocks.[17]

Officials thus recognised from the outset of CTBT negotiations that the safety and reliability of existing nuclear weapon stockpiles might present problems under a CTBT in force. London found it disturbing that it took the US so long to face up to the issue.[18] Following a UK request in September 1977, the State Department had provided a short paper setting out US views, which concluded that US weapons could be maintained 'with adequate confidence' without any nuclear explosions. Although this paper purported to be the considered and agreed views of the US, MOD and AWRE experts doubted that this statement was fully endorsed by all the relevant US agencies. For this reason, the UK did not accept its conclusions. MOD officials had also discovered that there was in fact no agreed inter-agency view at all on what was required to sustain stockpile safety and reliability.[19] It would seem that MOD and AWRE experts must have heard thus from their counterparts in the labs and DOE in the course of routine exchanges and contacts. US concerns only began to surface after a major Soviet concession in the negotiations on 2 November 1977 on acceptance of a moratorium on PNEs.[20] This concession might have been perceived as potentially increasing the prospects for a test ban, a possibility that exposed the disarray and divergences in the internal US policy process. If a treaty seemed far away, then there would be no need for any urgency to sort out negotiating positions and objectives. Moscow's concession proved a rude awakening. The other side might be serious and this might actually happen.

On 7 December 1977 at the tripartite negotiations in Geneva the US tabled a working paper without any prior warning to UK officials – something which no doubt would have generated a fair degree of angst given the lengths to which the British went to consult the US on all key issues. This working paper stated that the 'technology used to describe the substantive elements dealing with the prohibition of nuclear weapons tests and explosive devices intended for peaceful purposes may require mutually acceptable understandings regarding distinction between prohibited explosions and permitted nuclear experiments'.[21] Exactly what this would mean in practice was unclear to the UK, and the Russians for that matter. In further exchanges with DOD and DOE officials, UK experts learned that there was in fact no agreed view in Washington about the intention of this sentence and how the requirement might be met.[22] This was important since such experiments could impact the scope of the treaty and its verification. As noted, this was all done without any prior consultations with the UK and also on last-minute instructions from Washington. Paul Warnke, Head of the US delegation and of the Arms Control and Disarmament Agency (ACDA), could not provide either the UK or Soviet delegations with a satisfactory explanation.[23] GEN 63 agreed on 21 December that further technical discussions, particularly with the Americans were needed before a substantive report on the issues and options on permitted experiments could be prepared for Ministers.[24] A useful first step would be to send a senior team of officials to Washington to include Dr Robert Press (Cabinet Office) and Vic Macklen (DCA (PN), MOD) in order to gain further insights into US thinking.[25] However, officials worried that it might be difficult to persuade the US of the need for such a visit and it would be important to stress that the aim of such a meeting would not be to persuade them of any particular policy line, but simply to explore the technical aspects and to identify US thinking to date on the problem. GEN 63 saw three possible approaches open to the UK, the first and most difficult of which would be to attempt to draft precise technical definitions in the Treaty itself; the second would look for a separate understanding outside the Treaty, which would have to be confidential, but the existence of such a document would no doubt leak in due course; and third, agree to pass over the problem in silence. However, in this case, the CCD could raise this matter and discussions over definitions to be used in a treaty were bound to happen anyway.

As for definitions, and what might or might not be covered by the scope of a comprehensive test ban treaty, the MOD made clear that the UK would have to retain a capability to ensure the continued safety and reliability of its existing nuclear weapons stockpile, sustain the mutual benefit of the 1958 Mutual Defence Agreement, make and put into service warheads necessary to counter technological defence improvements introduced by the Soviet Union and to reflect changes in Allied nuclear policies.[26] The Chevaline Programme existed for this very reason. This capability had to be maintained despite the fact that the then-existing level of UK nuclear forces was set to a minimum and that its stock of tested warhead designs was small (and only two basic designs were in service the WE177 and ET317 with one design for Chevaline already tested and another lighter version planned for April 1978). For these reasons, the UK was not well equipped to react to changes in nuclear defence requirements under the restrictions that would apply under a CTBT. Instead, the UK would have to rely on theoretical and experimental work, hence the importance of permitted experiments. All UK warheads had components of US design as well as British manufactured ones, which gave an assured performance precisely because they were based on complementary US experience, hence the importance of the 1958 MDA and the UK's ability to undertake work that would convince the US that the 1958 agreement was worth continuing. Remanufacture of war-heads might also be necessary, for which experience and data from prior testing would be important to help validate the scientific and engineering judgements on reliability and safety. Exactly what sort of permitted experiments would be required in future under a CTBT to sustain all of this was unclear to officials. It was terra incognita. However, they could state confidently enough that five types of work would be essential for stockpile safety and reliability. These were firstly experiments with simulated fissile materials, including non-fissile pluto-nium and uranium isotopes, to investigate the assembly processes (i.e. how the nuclear explosive process progresses), which occur in nuclear warheads; second, small-scale trials in which high explosives operated on sub-critical quantities of plutonium in order to provide equation of state data;[27] third, pulsed experi-ments in which fusion reactions occurred, perhaps with some fission reactions, to validate in the laboratory, theoretical weapon physics (including inertial confinement fusion work); fourth, pulsed radiation experiments to check the vulnerability of equipment to nuclear weapon effects – these would use pulsed reactor and inertial confinement fusion systems; and finally, safety experiments that were not expected to give a nuclear yield, but which nonetheless involved a small risk of such a yield.[28] Similar issues arose during the 1958–1962 Con-ference on the Discontinuance of Nuclear Weapon Tests and this was precisely why the Maralinga Experimental Programme of minor trials was so important to the development of the UK nuclear weapons programme at that particular time.[29] At that stage, it had never been the intention to include such activities in a CTBT, and it looked as if that position would remain the same this time around too.

Come January 1978 there was still no agreed UK position on a definition of the experiments that should be permitted under a CTBT. Certainly, such experiments could not be verified and there would have been concerns about giving too much latitude to the Soviet Union to cheat to the West's disadvan-tage. Nevertheless, the UK thought that experiments should be permitted if they satisfied one of the following criteria: no neutrons were generated, the nuclear energy release occurred above ground in such a way as not to be inconsistent with obligations under existing treaties (the PTBT and TTBT), neutrons were generated for a period longer than 10 microseconds, or that the nuclear yield was less than 200 tons high explosive equivalent. Such criteria could, of course, not be verified under any conceivable verification regime that might be agreed under a treaty. Whether such technical language could actually be negotiated and agreed with the USSR was also a rather large unknown.

Ministers believed that the issue of permitted experiments had important technical and political implications that could complicate the CTBT negotiations, and as such the UK would require a fuller account of US intentions. The Prime Minister, Defence Secretary and Foreign Secretary therefore agreed to send a small team of senior officials to Washington for exploratory talks at the end of January 1978.[30,31] However, the US declared itself unable to accept such a visit as Washington would not be ready to field such a meeting for a month.[32] This inability of the US to sort itself out and agree on positions proved a recurring theme in these negotiations, and one moreover that imposed a brake on progress. As one US commentator observed, of all the issues which led to bureaucratic infighting in Washington, none prompted so vicious a brawl, for so long, as the CTBT negotiations.[33] But despite this, the Americans quickly came around to the idea and indicated that they could accept a visit on 16–17 February 1978 and asked, interestingly enough, for a paper on the UK's past experience on nuclear weapon stockpile maintenance.[34] Clearly then it was not a matter of one-way traffic: the UK experience had potentially much to offer the US weapons labs. We can also see at this time that the US Joint Chiefs of Staff (JCS) were adamantly opposed to a CTBT. They told the President firmly that a treaty was not in the US national interest, and warned him about a treaty's adverse impact on the reliability of the US deterrent.[35] US and UK weapons design staff had separately concluded that the data on which the experience of nuclear warhead designers was based, and on which the continuing capability to support safety and reliability assessment depended, could not be derived from laboratory experiments alone, but would require occasional nuclear tests of 2–3 kilotons.[36] It was also important to retain the competent experience upon which critical judgements would have to be made.[37] This would be the underlying concern; it was not so much testing per se that was important, but rather the ability to retain the expertise needed to make sound assessments on the reliability and safety of existing weapons, an expertise that could only be sustained through periodic testing. When test results matched the predictions, then that when repeated over time, reinforced confidence levels in any scientific or engineering judgements that might have to be made on safety and reliability.

An added concern was that the then limitation in verification capabilities, and what would be negotiable with the Soviet Union given the closed nature of its society, meant that the USSR might get away with testing at up to ten kilotons. Testing at these levels would not only allow the USSR to maintain its nuclear capability, but it could also permit new nuclear weapon development to occur.[38] This in part explains why the UK wanted both sides to be able to conduct a small number of very low-yield tests so that the West could maintain the effectiveness of its own nuclear weapon stockpiles. Thus, even this early in the process, the UK was continuing to move away from a truly comprehensive treaty.

UK–US technical exchanges: February 1978

AWRE gave a presentation at the February Washington meeting on 16–17 February 1978 on UK stockpile reliability without major nuclear testing. This was an area in which the US showed great interest.[39] Vic Macklen told the US that UK experience from the late 1960s through to the early 1970s had shown that even the modest design changes to the Polaris warhead still left both UK and US designers uncertain as to acceptance of the design into service without testing it first. In terms of stockpile reconstruction and weapon development, the gap in weapon-device testing between 1965 and 1974 was not as definitive for the UK as it might appear, particularly since the UK was exchanging views and obtaining advice from the US laboratories that were carrying out a great number of tests during the period. If it had proved essential, for stockpile reasons for example, the UK could have tested at any time without any treaty constraints.[40] AWRE's advice was that an ability to conduct tests in the upper end of the range of one to three kilotons was required for firm assurances on the safety, reliability and credibility of the UK stockpile. If they were limited to tests of 100 tons, work would not be relevant to ET 317 and WE177 weapons and assurances would not be valid for more than about five years. By that time the present generation of weapon designers would have retired. A limit of three kilotons would not allow more than a very limited opportunity for the development of new designs; tests in the 10–15 kiloton bracket would be required to allow vigorous development. Moreover, without a continued ability to test at three kilotons, it would not be possible to put new 'super-safe' explosives into any of the UK stockpiled weapons.

US interests here had originally stemmed from a request by John Marcus of the National Security Council (NSC) for a paper setting out the UK experience with stockpile reliability given the fact that the UK had not tested between 1965 and 1974.[41] As Denis Fakley of Defence Science, MOD had told the Americans in March 1977 before the tripartite negotiations had even started, the UK had found it extremely difficult to maintain the competence of its weapons laboratories in the absence of testing and had to rely on a surveillance programme to maintain the viability of the UK stockpile as well as US advice provided under the 1958 MDA. The UK, however, had never tested for stockpile reliability but would have done so had a need emerged.[42] The UK's capability was thus lower than it would have been had it continued to test, which had been the United Kingdom Atomic Energy Authority's (UKAEA) intention back in 1965 with its plans for a generous research programme of nuclear tests.[43] In terms of the US request, the MOD, however, wanted to ensure that any such paper was presented to all the US agencies at the same time.[44] Aldermaston had been attempting to assess what effect a cessation of nuclear testing would have on the reliability of the stockpile in the future and the bearing on this of the UK's past experience of how its stockpile behaved under the arrangements implemented for assuring safety and serviceability.[45] Peter Jones, Chief of Warhead Development at AWRE, made several major points on the British experience in his detailed presentation during the February meeting. First, the Polaris warhead was re-designed primarily for nuclear hardening reasons stemming from the Soviet deployment of ABMs, which meant that the Polaris ET317 warheads would all be replaced before deterioration could become a problem. Second, AWRE was studying what could be done about the WE177 bombs jointly with the RAF and Royal Navy. Even if it were to be concluded that no military requirement to improve the type or performance of the warhead, the recent testing experience for Chevaline made two important contributions: it had produced a design compatible with the WE177 if the existing life span of the weapon gave time to apply them – the weapon had a design life of 18 years, and it re-vitalised the UK's capability to assess warhead deterioration. Prediction of the effects of deterioration in the bomb – and there were some – was helped by the theoretical experts who had acquired much of their know-how by their involvement in the Chevaline programme. Some aspects of deterioration required assessment through engineering judgement and this was adversely affected by the movement of key 1965 development experts out of this field, or by the imprecise recollection of those still in post. During the eight and half year gap in UK testing AWRE had high confidence in the reliability of its relatively new weapons, it did not follow that it would have remained so had Aldermaston not acquired new experience from testing when it resumed in 1974. UK weapons had not been designed with an expectation that future testing would be prohibited. Third, it was important to note that changes in Soviet non-nuclear capabilities could, and did have, a marked effect on UK forces. Indeed it was the development of radars and interceptor missiles that forced the UK to undertake the Chevaline modifications of the Polaris payload, including its warheads. Fourth, UK weapons were particularly vulnerable to deterioration problems given the fact that it was using only two designs with a lot of common features between them. Fifth, the UK only tested infrequently because of the few systems that it deployed and resource limitations. Tests were conducted because the UK really needed to, not just for the luxury of scientific interest. A further consequence was that the UK was very restrictive on empirical development through nuclear testing and Aldermaston mustered the best theoretical prediction and laboratory evidence that it could before committing to a test. Without this R&D capability, Aldermaston would lack the basis of judgement for an assessment, both of the stockpile and of variation in production skills and processes, with the consequence that assurance of stockpile reliability would suffer.

Aldermaston used several approaches to assess stockpile reliability by gathering evidence on the condition of the warheads' nuclear assembly and then making an assessment of whether the departures from the 'as-designed' condition would be such as to affect performance. It would only be in this context that testing would be relevant. Methods deployed by AWRE included: retention of weapons materials samples for accelerated testing in the laboratory; placement of nuclear safe 'surveillance rounds' in the stockpile; routine withdrawal and breakdown of service warheads; breakdown of service warheads for refurbishment and trickle production of new ones; examination of obsolete warheads and assessment of the defects and prediction of their effect on weapon performance. Surveillance activities were aimed at providing advance warning of potential stockpile problems by applying environments to warheads under test that seemed likely to accelerate degradation. Higher temperatures could, for example, accelerate corrosion or other chemical effects on the diverse materials used in a warhead such as adhesives, or the vibration seen in aircraft carriage could be applied more frequently than service rounds were likely to see. Surveillance rounds were subjected to all environments seen in service and in flying these AWRE avoided the Ministerial prohibitions that applied to the airborne carriage by aircraft of live weapons in peacetime. Aldermaston maintained a positive programme for the withdrawal of normal service stockpile weapons, about one per year for each type, for breakdown and examination during the in-service downstream life. These were replaced into the stockpile by drawing on the spares of non-operational warheads, but once production had finished it was necessary either to have stocks of component parts for rebuild as required, or such components then had to be specially made at the time they were required. This was one of the circumstances that drove the UK to adopt the 'Trickle philosophy'.[46] If the UK were prevented from testing, it would probably have to revert to trickle production in order to retain its ability to produce weapons. A designer's capability to judge unexpected deterioration in the stockpile depended on his experience with testing. For this reason, UK experts believed that if they could test up to 1–3 kilotons, they could retain their basic capability to make judgements about the stockpile, but without testing these would erode after five years and could not be replaced by staff even if they were given scientifically interesting work.[47] Perhaps this was an over-pessimistic view of staff motivations, but in defence planning worst-case scenarios do have a tendency to predominate. However, these methods under trickle production only provided the data – conclusions depended vitally on the capability of AWRE R&D experts to assess the effects on performance, and therefore largely on the experience they, or their successors, acquired since the weapons entered operational service. Several specific effects had been observed – changes in detonator bridge wires, relaxations developing in plastic structural components, corrosion in nuclear material and migration of corrosion products to regions that should be clear of them. In no case had the likely effects of these developments exceeded AWRE's understanding that a full nuclear test was required to evaluate them, and this experience implied that they were unlikely to do so in future. Moreover, this process had not revealed any serious degradation problem that would have required a premature rebuild of the stockpile. The crucial factor in all of this was the capability of the R&D experts at Aldermaston and a future lack of test experience affected this capability; it was only since 1974, when the UK resumed testing, that the scientists and engineers fully appreciated how much better their technical understanding could be and how many staff lacked this important constituent of judgement.

Some of the key points that emerged during subsequent US and UK discussions in London were summarised by Ken Johnston (Defence Science, MOD) for Bob Press in the Cabinet Office.[48] Peter Jones had noted that in reviewing UK requirements, a yield limit of three to four kilotons, although restrictive, would permit maintenance of capability and expertise, but allow only limited scope for development – the Americans agreed. As for the numbers of tests, the UK's needs could be met with an average of up to one test a year. In the face of a complete ban that would not be implemented until a period of two years had elapsed, Jones noted in response to a US question that resource constraints would preclude a large number of British tests. However, it might be worth considering testing a device with insensitive explosive and possibly one with a plutonium-free trigger on the grounds of safety, while any improvements would be aimed at a warhead for a Polaris successor system. It is worth noting that this was long before any political decision had been taken on whether to build a successor Polaris. AWRE and MOD were clearly about keeping options open, not foreclosing them. Vic Macklen (DCA (PN) MOD) added that it might also be worth conducting tests aimed at discovering how useful a permitted yield limit might be. In sharp contrast, the US noted that they would conduct between 40 and 50 tests after a year's preparation. On the subject of the verifiability of a three-kiloton threshold, UK and American views diverged. The UK thought that three kilotons lay below the detection limit of what was then achievable by seismic means whereas US views were mixed, some expressing the view that this could be done reliably whilst others were more doubtful. However, both agreed that a three-kiloton threshold would pose significant challenges for a non-nuclear weapon state conducting its first test since it would find it hard to be sure that the yield would keep within that limit.

GEN 63 reviewed the outcome of the Washington visit and subsequent meetings in London on 23 February.[49] The main points were clear: some nuclear tests with yields up to three kilotons would be necessary to ensure the safety and reliability of nuclear weapons stockpiles, though such tests would not be sufficient to allow the development of new weapon techniques, for which tests with yields of at least ten kilotons would be required. The significance of this threshold was that was the very one which was beyond the capabilities of a seismic verification system to detect and identify as assessed by AWRE Blacknest's seismic experts. Officials assessed that given the closed nature of Soviet society, the USSR might be able to get away with testing at ten kilotons, which would enable it not only to maintain the safety and reliability of its stockpile, but to develop new weapons.[50] There was later some doubt about this assessment, but we are getting ahead of ourselves. US officials claimed that new improvements in verification technology would be available by the time the treaty was in force that would permit the detection and verification of underground nuclear explosions with yields of three kilotons and even in some cases of verifying explosions of only about 0.5 kilotons.

UK officials thought that these meetings with US experts had been successful since they had established clearly that the point worrying the US, and on which there was no inter-agency dispute, was that tests would continue to be necessary for the safety and reliability of current stockpiles. AWRE experts agreed with the US that tests of up to three kilotons were needed for safety and reliability and that yields of ten were needed for new designs, but not inconceivable that an ingenious scientist could find a way to achieve such development within a three-kiloton limit. Now, these conclusions would be difficult to reconcile with HMG's public political commitment to an unlimited test ban treaty, and that of course was the dilemma that the UK had to grapple with for the remainder of 1978, and indeed beyond. The key question facing Ministers and officials now was how to reconcile the apparently diametrically opposed policy require-ments, which of course was hardly the first time that this problem had faced UK nuclear defence and arms control policy. It was very much a case of 'déjà-vu all over again' with clear echoes of the late 1950s.

So what could be done about this? Britain had a few options: ignore the problem and continue negotiations without further reference to definitions and permitted experiments, but knowing what they now did, this course of action was now not open to either the UK or US. The US at first had thought that the problem was one of permitted experiments, but had come to realise that the difficulties did not concern permitted experiments such as laser fusion, but the military question of stockpile maintenance.[51] A treaty Review Conference could be held after five years to consider inter alia the need for the nuclear weapons states to resume testing for safety and reliability reasons. This had some advantages: it would deal with the problem and thereby maintain national security, but even a gap of five years without full testing might undermine confidence in stockpile reliability and safety. Officials thought that Ministers might accept such a risk for test ban treaty, but the armed forces might grow concerned over nagging doubts over the safety and reliability of the nuclear weapons under their charge. Another possibility would be to set a threshold of permitted tests in treaty as well an annual quota for each weapon state. AWRE thought that one nuclear test a year would be enough to meet UK require-ments, although it would be likely that the Americans might want five or six. Now whilst this would be an improvement over the 1974 Threshold Test ban Treaty (TTBT), it was hardly likely to satisfy demands for a CTBT and would be open to strong criticism from the non-nuclear weapons states (NNWS).[52] Extreme conservatism in the weapons communities in the labs and nuclear defence policy officials on both sides of the Atlantic seems to have characterised thinking about a test ban at this crucial juncture – if in doubt err on the side of caution, which in the case of nuclear weapons safety was probably no bad thing.

GEN 63 concluded that provision in the treaty would have to be made for tests up to three kilotons, but how could this be squared with HMG's very public political commitment to a CTBT? Officials knew that they could not recommend to Ministers that a truly comprehensive treaty could be accepted as this would jeopardise UK national security. The option of relying on a Review Conference could not, on balance, be recommended to Ministers because of safety considerations, although the debate in GEN seems at that stage to have focused more on the reliability side of the equation. This left the option of building a three-kiloton threshold into the CTBT, which could create severe political problems as such a treaty would not be comprehensive, but which seemed to be the only option available given what AWRE knew of the technical challenges. There would be other issues too in terms of the ability of any seismic verification system to determine that a test was at, or above this threshold, but that is another perhaps even more complex issue given what was to come with the US consistently over-estimating the yield of Soviet tests at the Semipalatinsk Test Site.[53] UK estimates of Soviet test yields were consistently lower than those of the US, and following a reassessment of the UK seismic calculation for the test site in question, the revised estimates for all Soviet tests since 1976 gave no yields greater than 150 kilotons.[54] Meanwhile, the next step called for the experts to prepare a report for Ministers. MOD and FCO would also seek urgent confirmation of the American claims on verification capabilities. MOD was tasked to prepare a paper for the Defence Secretary to circulate to key Cabinet colleagues setting out essential defence requirements for maintaining the safety and reliability of the present UK nuclear weapon stockpile.

The Foreign Secretary's concerns

All of the above provides the context for the consequent Ministerial exchanges on the stockpile reliability issue and in particular Dr David Owen's views, stoked up by Lord Zuckerman's critical assessment of the arguments aired by the nuclear weapons community in MOD Main Building, the Cabinet Office and down Aldermaston way; and, of course, across the pond in the US.[55] Solly Zuckerman had now no formal role in nuclear policy-making as he had ceased to be government Chief Scientific Adviser in 1971, though he retained an office in the Cabinet Office.[56] Despite this diminished official status, he nonetheless was not shy about making his views known.

Dr Owen had major problems with these developments in US testing policy and told the Prime Minister at the end of March 1978 that he felt that consideration of the stockpile issue should start from the major point that the UK was politically committed to a CTBT. He reminded Callaghan that in his statement to the House of Commons on 17 June 1977 about the decision to start the current tripartite negotiations he had referred throughout to a comprehensive ban. Therefore, any proposal that derogated from the comprehensive nature of the ban would be a departure from the government's very public commitment.[57] Owen argued that there was also the important question of non-proliferation. A test ban treaty's value in this respect depended upon the adherence of a few important non-nuclear weapon states outside the NPT, and the most import of these was India which would be highly unlikely to sign up to anything less than a truly comprehensive test ban treaty. In addition, if it became known that the UK was only interested in an incomplete treaty, then this would cause all sorts of political problems for the UK at the forthcoming UN Special Session on Disarmament in New York.

Vic Macklen in MOD was not impressed by these arguments; for him, a treaty limited to five years was nothing more than a moratorium. At the end of the day, it would be the US that would decide. Given the UK's dependence on the US, it would not help to put one-sided political viewpoints that would have no impact on Congress.[58] The UK could not guarantee that there would be no pressure to continue with a comprehensive test ban after five years, and there would likely be no sympathy from the non-nuclear weapons states parties to such a treaty for the UK and US if either wanted to make some allowance for stockpile maintenance testing. Macklen felt that the political costs of resuming testing at this point would be far greater than opting for a threshold/quota treaty at that stage of the negotiations. AWRE, moreover, was not able to guarantee future safety and reliability by testing at that point in time as there were no per-ceived stockpile issues that would warrant such a requirement, but that might not be true in future years. In any case, the UK could not just test immediately as it took more than a year to plan, prepare and actually conduct a test.[59]

Sir Solly Zuckerman had reviewed the GEN 63 (78) 4 paper on the stockpile issue and submitted his views on it separately to Dr Owen in a personal minute – to say that he was sceptical of the technical arguments would be an understate-ment and his line was much the same as it had been in the 1960s when he was Chief Scientific Advisor to the government.[60] Zuckerman argued that it was neither the US nor USSR that had first raised this matter of reliability, but the UK itself. He noted that the MOD had not accepted the US September 1977 official assessment, and that dissatisfied with this answer 'our technical people approached their opposite numbers in the US weapons laboratories, from whom they elicited an echo of their own doubts'. Zuckerman doubted the arguments on the need for testing to sustain reliability and the conclusions that could be drawn for the stockpile as a whole. He also wanted to know what reliability and safety testing had been required before now and why guarantees could not be given for more than three to five years. These views were to be echoed later in Owen's minute to the Prime Minister. Once he found out about the source of Owen's information, Macklen was dismissive since he commented that Zuckerman was not a nuclear warhead designer and that his own view on stockpile maintenance was not likely to be founded on fact or modern tech-nology; and most, if not all of his US contacts, were in the same boat.[61] There is a view here that maintains that nuclear weapons knowledge is confined to a handful of individuals and that unless you were one of those individuals, your opinion would not count. That was clearly Macklen's view and a reality that made it extremely difficult for Ministers to gainsay advice from such individuals.

Owen remained convinced that a Treaty had to be comprehensive if it were to be really useful as an arms control and non-proliferation instrument. This had to exclude any provision for continued testing, whether in a 'qualified national interest clause' or quota cum threshold idea. He did not much care for a Review Conference since this would require it to be made clear at the time of conclusion of the treaty that stockpile maintenance and resumed testing would be discussed at such a conference – such a prospect would only deter the NNWS from joining such a treaty in the first place. It was also clear to Owen that, judging by the technical paper attached to Fred Mulley's minute on the issue, the problems observed so far in the UK warhead stockpile had always been rectified without the need for a test. As noted, there were no UK tests between 1965 and 1974 and the tests since then had been connected to Chevaline development and other research goals – and certainly no tests from 1956 to 1965 were connected in any way to stockpile reliability issues that had emerged through warhead surveillance programmes.[62] That said, warheads were not designed with very long shelf-lives in mind at this period. Since it appeared that MOD felt that the present ability of AWRE design personnel to judge stockpile problems would last for about five to seven years, this would suggest, in Owen's view, a solution to the problem of stockpile reliability, maintaining a comprehensive test ban treaty objective and sustaining the non-proliferation goals of such a treaty. A CTBT of an initial limited period of five years would, in Owen's view, square the circle. Moreover, a decision to opt for a five-year treaty – that would also include a ban on PNEs, which the Russians had wanted to exclude – could be presented and defended publicly to the NNWS. The USSR wanted a three-year-long treaty that would cease to exist if France and China did not accede before the end of that period. There would, however, still need to be a Review Conference at the fifth year to decide whether the Treaty should lapse, be renewed or replaced. Owen did not buy some of the technical arguments made by the MOD. He observed that most of the UK's present warheads dated from 1965 to 1968 – in other words, ten to 12 years old and it was odd that no tests for safety and reliability of the kind now proposed had been required before the UK embarked on test ban treaty negotiations.[63] He was obviously unaware of the sort of arguments that Peter Jones had aired as noted earlier in this chapter in response to US questions on the UK experience with its self-imposed testing moratorium.

MOD was, unsurprisingly, not fully convinced by Owen's arguments, and as one official noted, 'a number of the points the Foreign and Commonwealth Secretary seeks to make in his minute are, however, susceptible to argument',[64] a polite way of saying he was plain wrong. MOD saw dangers in the five-year treaty option. First, there would be degradation in the capability of AWRE's design staff and there would be no provision for resolving an unexpected problem in the stockpile which might arise. But second and most important was the question of what would happen at the end of the five years. MOD felt that the UK could not be confident that it could extricate itself from the Treaty should it wish to do so without severe political penalties and these would be greater than those that might arise stage if a decision were taken at that point in time to go for a low threshold/quota treaty. Macklen had argued that the passage of only five to seven years would be sufficient to cause the loss of the ability to reach authoritative judgements, not that the ability to make such judgements would last for this period. For MOD there were likely to be enormous pressures to continue with the Treaty after five years. There would need to be an assurance at the outset that stockpile problems could be covered at a review conference at the end of the five-year period as the current treaty language on this topic only talked about 'review of the operation of this treaty . . . with a view to assuring that . . . the provisions of the Treaty are being fully realised' would explicitly exclude such factors. The Russians could no doubt be counted upon to take this line. However, the MOD saw dangers in prior notification that stockpile issues would be considered as part of this review. If the UK could not secure support for such a provision, a likely outcome in the MOD view, the UK would then be forced to cause the Treaty's collapse either by opting out or withholding support for its continuation. Such an argument would certainly strike a chord with Ministers as they would be anxious to avoid being placed in a position where the UK would be not only isolated, but blamed domestically and internationally for wreaking the treaty.

MOD conceded that it was indeed true that the UK had conducted no tests specifically for the purpose of safety and reliability. There had been no insuperable problems in the first decade of the existing warheads' life, but the second decade now beginning could be more problematical. More importantly, although the UK tests had been required primarily for improvement programmes, AWRE aimed to extract as much information from each test. None of the tests since 1964 has had a single or simple objective. The nuclear tests themselves were the culmination and proof of a long series of non-nuclear testing that enabled AWRE to exercise the expertise of the design teams and provide reasonable proof of its continued validity. There was a major surveillance programme for in-service warheads.[65] Aldermaston was unable by testing at that stage to guarantee future safety and reliability. Such tests were not needed because there were no perceived stockpile problems acute enough to require testing. It simply was not possible to foresee all possible problems that might arise with the sorts of issues that would come in their wake. And if the Foreign Secretary was proposing that the UK should set up a series of tests, these could not be fitted into the current test schedule as it took more than a year to plan and conduct a test as already noted. Crucially for MOD, the key was the need to satisfy its own experts that any solution the Americans might adopt to the problem of stockpile safety and reliability would be applicable in the UK. This seems a little odd given the very great level of dependence that the UK had on US technological assistance to maintain its programme – if it worked for the US then surely it ought to work for the UK? Nevertheless as we saw in the early 1960s, when the UK was 'anglicising' US designs there were considerable difficulties, not least with securing Ordnance Board approval for new designs to enter service.[66] So on balance there appears to be a solid basis for the MOD requirement that UK experts should be satisfied with US approaches to reliability and safety issues – they would also have to work in the UK with its own standards and practices.

Competing priorities made it difficult to arrange a meeting to review and decide upon British policy of Ministers before Easter 1978.[67] A key driver here was that President Carter was expected to make a final decision on this matter in March. Meanwhile, the preparation of briefing papers setting out the issues and options for the Prime Minister, Defence Secretary and Foreign Secretary continued. GEN 63 felt that it was now important not to make a distinction between safety and reliability as both were equally important. The need to guarantee the safety of nuclear warheads, officials thought, would be likely to be regarded as a more important consideration by British Ministers. Therefore, both arguments needed to be brought out in papers for Ministers, but the safety argument should be given priority. Perhaps officials were stressing this as a reason to avoid committing to a comprehensive test ban and by playing this factor up they could be sure that Ministers would have no option but to accept their advice. It would take a courageous Prime Minister indeed to disregard advice on the continuing safety of British nuclear weapons. This is perhaps harsh as AWRE's technical arguments were genuine – with maturing stockpiles AWRE could not be absolutely sure of their condition in say 1984 or 1985 assuming a five-year treaty had entered into force in 1979 or 1980. By that time the RAF's high-yield WE177B, for example, would have been in service for 23/24 years – it had a design life of 18, whilst the Royal Navy's WE177As had been in service since January 1969.[68] There might however be fewer concerns about the more recently introduced WE177C and the new Chevaline warhead design. Officials were no nearer solving what they saw as the major problem; namely, finding a way to square the circle of competing policy requirements. The obvious solution on reliability and safety would be unacceptable to the NNWS and would not further UK non-proliferation objectives. Of the various options, the only one that might prove tolerable diplomatically was to include warhead safety and reliability issues on the agenda of a five-year review conference and for emergency procedures for dealing with any safety problems in the interim. Officials saw no change in the need for assurance that Aldermaston's designers would be able to recognise whether or not a problem had developed and that any proposals for rectifying faults were technically sound. Their judgement could only be maintained at a sufficiently high standard if they had access to nuclear tests every two or three years.[69]

Vic Macklen, DC (PN), MOD set out the detailed case setting out the problems for the UK on stockpile reliability and safety in a paper intended for Ministers on 13 March 1978.[70] Macklen's arguments essentially underpinned the British position. UK warheads in service were designed to have a service life of 16–20 years, and as such categoric assurances on safety and reliability could not be given to cover the entirety of a warhead's future lifetime. Furthermore, since the UK had never been able to carry out a large number of nuclear tests, confidence in the reliability and safety of the nuclear warhead rested on a meticulous examination of its components plus a judgement of the results of this examination by a design team whose experience had been based on actual nuclear testing of experimental devices. UK weapons produced in the late

1960s and 1970s – Polaris ET317 and WE177 – were smaller and lighter than the first- and second-generation UK weapons – Blue Danube, Red Beard and Red Snow – as they were based on much more refined design principles, which permitted little latitude for variation outside the specified engineering toler- ances. As stockpile weapons aged, the surveillance programmes occasionally revealed the incidence of such effects as corrosion, cracking, generation of gases which could produce explosive mixtures, powdering of explosive components for example, within the nuclear warhead. All of these would pose problems for the weapons designers who had to be able to give dependable assessments of the effects on safety and reliability. However, there was no guarantee against some future defect falling outside the experience of the design teams. Unfortunately, the policy of rebuilding weapons to exact copies of the original design would become over time impracticable to carry out without recourse to nuclear testing because of the variations in materials and non-availability of original components. To validate any rebuild, practical nuclear warhead design and nuclear testing experience were deemed essential.

The essential data on which the experience of nuclear warhead designers was based were those obtained by nuclear testing of the actual device produced by the designers. For the UK, such tests had been very few in number, which was because it had access to some US design data through the 1958 MDA. With- out such data, the number of tests needed to develop the current UK nuclear stockpile would have been much greater. However, this benefit also carried a down side too in that the small number of UK tests meant that there was little evidence available on the sensitivity of the existing warhead designs to small changes in dimensions, or materials specification. A further problem and, as noted earlier, was one related to the fact that the UK only had two basic types of warhead, was that a serious problem concerning a given type would affect a large proportion of the stockpile and could not be cured immediately by select- ing a suitable existing alternative warhead. Under a CTBT, the availability of designers whose concepts and judgements had actually been submitted to the rigour of nuclear testing was expected to decrease quite rapidly, thus degrading post-design support capability. For AWRE, given the small number of person- nel involved and their age structure, the passage of five to seven years would be sufficient to cause the loss of the ability to reach authoritative judgements across the broad range of possible stockpile problems. A test ban would also prevent the replacement of the current high explosives used in current warheads with an extremely insensitive version, which would be much safer and useful for tactical weapons as if these were ever captured by terrorists, the weapons would be virtually impossible to detonate. But given the nature of the new explosives, this development would need to be validated by nuclear testing. For these reasons, it was the united technical view of those responsible in the UK for arriving at judgements on stockpile safety and reliability that permitted experiments such as hydrodynamic testing would not provide sufficient experi- ence to allow design teams indefinitely to underwrite the safety and reliability of the stockpile.[71] The only option open to address this requirement would be by conducting occasional nuclear tests with yields of 2–3 kilotons at a frequency for the UK not greater than an average of one per year. In AWRE's view, testing in the region of 3 kilotons would not permit the development of novel strategic nuclear warheads, though that said such a yield could be used by weapons states to develop new low-yield warheads for battlefield weapons.

In fact, in some quarters in Washington, it had struck one US official – John Marcum (NSC) – that concern in the UK about the disadvantages of a CTBT seemed to predominate over support for the treaty. Marcum thought that the FCO was virtually neutral, whereas the line in the MOD and the Cabinet Office was clearly hostile. Needless to say, the Washington Embassy thought that this was hardly a fair description of the FCO position, though it is instructive to note that the Embassy did not say it was a misreading of the MOD or Cabinet Office view.[72] Despite all this, the London view was that Ministers had all the arguments in mind, not just the stockpile ones.[73] There was clearly some disagreement in GEN 63 over the options for next steps, although not stated explicitly in the notes of the meetings it is possible that there were divergences, or at least differing degrees of emphasis on the options and problems between the Committee's FCO and MOD members.[74] GEN 63 was thus unable to agree on which solutions to recommend to Ministers, but it was clear that the UK could not make further progress without knowing what the US position was, especially on the political implications and for this reason, further high-level official discussions were deemed desirable – hence the need for yet a further approach in Washington.[75] Working hand-in-glove with the US would thus be essential for the UK and consequently the Cabinet Secretary, John Hunt, duly advised the Prime Minister that HMA Washington should speak to Dr Zbigniew Brzezinski to register UK interest and propose high-level bilateral consultations.[76] Such a message would also underline that the UK was examining the warhead safety and reliability issue under a CTBT and hoped that the US would not come to any conclusions before holding further discussions with the UK, and of the UK's wish to be closely associated with US consideration of the problem.

Ministers eventually found time to consider the safety and reliability problem when the Cabinet sub-committee on nuclear defence policy discussed the matter on 3 April 1978.[77] At the end of the discussions of the options, the Prime Minister summed up the feeling of the meeting. Callaghan noted that President Carter had told him during his recent visit to Washington that he was anxious to secure an agreement on CTBT before July. Britain was also committed to a test ban, a point that he had made clear to the Indian Prime Minister in January. Ministers agreed that the UK should not propose exemptions (low-yield tests/quotas) to the test ban, but that the most satisfactory compromise would be to pursue a treaty of limited duration, on the lines proposed by the Foreign Secretary. This should be for five years, but, if necessary, a shorter period could be accepted too. All options would be kept open thereafter – a new treaty would be negotiated, the current one would be extended for a further time-bound period or indefinitely or a treaty with a threshold and small and diminishing quotas. Consequently, HMA Washington would be instructed to inform the US of this conclusion and both Foreign and Defence Secretaries would take an early opportunity to explain UK thinking to their US counterparts. FCO officials noted that this new requirement was to provide the possibility that, sometime after entry into force of the treaty, the US and or the UK might decide that it was essential to conduct one or more nuclear tests in order to ensure the reliability and/or safety of their weapon stockpiles.[78] The Prime Minister himself took the view that if small explosions were to be allowed whilst the Treaty was in force, its credibility would be greatly reduced.[79] One might go further and suggest that its value as a non-proliferation measure would be zero.

Callaghan duly spoke to Carter by telephone on 17 April 1978 in which he noted that US and UK experts were now saying that 'we will need a few controlled explosions once the Treaty has gone into effect, in order to verify existing stockpiles. I hope you will look into this. I would need a lot of convincing that it made sense'. The President promised to examine the technical issues and respond if he found something more.[80] British Ambassador Peter Jay followed this up with a conversation with Brzezinski on 21 April and asked him how the US intended to handle the problem of maintaining the reliability of nuclear stockpiles under a CTBT. Brzezinski indicated that he did not have much to say at present and that the US was looking at three options: a five-year review procedure, an exemption (i.e. a Threshold Test Ban Treaty), and one other, which he did not specify. He also indicated that with some of these options, there in effect was no CTBT treaty; and that the US first had to know how severe the stockpile degradation problem was before determining whether to choose an option that was less politically significant. In response to Jay's question about the value of a CTBT on vertical (within the existing nuclear weapon states) and horizontal proliferation (to non-nuclear weapon states), Brzezinski indicated that there was a subjective judgement here; but that he saw the importance of such a treaty as being more in terms of US-Soviet relations than proliferation. This is clearly a major difference from the UK where non-proliferation was perhaps the main reason for such a treaty. Jay indicated the Prime Minister's concern about a CTBT in terms of the Indian nuclear problem to which Brzezinski said that the US was not approaching this issue as an excuse to back out of a CTBT – although there 'were some people in the government who would use the issue in this way' – but rather because there is a serious problem for review. He promised Jay that the US would move quickly, and within two weeks they would be giving it close attention. Given the 4 May resumption of the tripartite negotiations in Geneva, he suggested that they should have a meeting on this question in the following week. There were a large number of unknowns – including on US (and by implication UK) needs. Jay asked that the Administration should understand the UK's position on this.[81] The PM clearly wanted a comprehensive test ban treaty, but was also worried about the safety issue. Finding a way to cut the Gordian Knot was proving elusive.

Ministers decided that the UK should not appear to take sides in the internal US debate ahead of the President's decision on the reliability, safety and permitted experiments problem. It would no doubt not do to end up on the losing side, which could have adverse implications for UK interests. GEN 63 was, nevertheless, fully confident that President Carter would consult the Prime Minister before taking a decision which differed from UK views. A withdrawal clause in the Treaty opened up some possibilities since the UK could come out of the Treaty and test if it really had to; moreover, it was not unusual to have such a clause in arms control treaties as the PTBT 1963, Nuclear Non-Proliferation Treaty (NPT) 1968 and Biological and Toxin Weapon Convention (BTWC) 1972 all had such provisions – 'the supreme national interest' clause. There would however be a problem in having to explain in detail to the United Nations just why the UK had to withdraw for stockpile reliability and safety reasons – indeed there might be very little that could actually be said about such reasons for non-proliferation (revealing warhead design details) and security reasons (exposing a vulnerability in the deterrent). In any case, it would not be a question of the UK withdrawing from a CTBT to conduct a test unless the US also withdrew since the UK would have to test at Nevada.[82] And in any case, the UK could not take a definitive view on the question of a provision of an explanation of withdrawal until it had discussed the problem with the US. Use of the word 'initially' in a treaty of limited duration might be a deterrent to some NNWS as it implied that the treaty would not be automatically extendable; and the Review Conference option might be problematic too as the current draft under consideration in Geneva only talked about a review of the treaty's operation and whether it should remain in force. FCO officials had prepared a series of draft Articles for the treaty that reflected Ministers' views on duration and these were considered by GEN 63, but the crucial one on five years' duration was to be kept back for the time being.

Meanwhile, the US was still considering the options and the UK's five-year compromise was only one idea in the melting pot.[83] Annual quotas were also under consideration and there were three variants here which entailed unlimited testing with yields below a stated threshold; an agreed quota below a set threshold and an agreed quota, but with a diminishing yield down to zero. Any quota system for the UK would present an unusual problem since an apparent minimum of one test per year would be a considerably higher rate of testing by the UK than for the previous decade. A yield threshold would have to be carefully designed as would the period of diminution. Negotiating such a variant with the USSR would be very difficult, but the advantages of this option were that it would permit testing to solve any safety and reliability problems that might arise. In addition, if combined with a suitably low-yield threshold, this option would also prevent the development of a new generation of nuclear weapons and testing by any NNWS. A diminishing quota would lead to a test ban in due course, though the technical basis for assuming that there would be no stockpile reliability problems had not been substantiated. However, thresholds would not appeal to NNWS and non-NPT states as they would undoubtedly see this as a discriminatory measure in favour of the existing nuclear weapon states; and, if this approach were taken, the UK and the US could now hardly object to the Soviet desire to exempt PNEs from the scope of any ban. Thus far the UK had argued that PNEs could be used clandestinely to further improve weapons design. As for the next steps, the UK could still not really move until Carter had made up his mind – this was to be the pivotal event in the whole tripartite test ban negotiations – and until then the UK resolved to take no action that might be construed as an attempt to influence the debate among the various squabbling US agencies. As Percy Craddock was to observe, it was now abundantly clear that the tensions between security (stockpile safety and reliability) and political considerations (non-proliferation and halting the arms race) were sharper than first supposed when the treaty negotiation had begun back in the summer of 1977.[84]

Carter decides

President Carter decided in May 1978 to opt for a treaty of five years' duration – essentially the British preference, which Ministers had concluded at their 3 April meeting to be the most satisfactory compromise under the circumstances.[85] The essence of Carter's decision was that the Treaty would automatically terminate at the end of five years with no legal commitment to extend, and with a review conference during the final year to decide whether it should be replaced by another treaty. Moreover, the Carter Administration intended to tell the Senate openly that it would plan to resume testing for safety and reliability purposes unless by that time such tests were not in fact necessary.[86]

Brzezinski, Carter's National Security Adviser, had laid out the options in a memorandum for the President on 10 May that made clear the reluctance of the weapons labs directors and the Joint Chiefs to accept an indefinite duration test ban; they would have preferred a treaty lasting three, but could just about live with a five-year ban provided there was an express commitment to resume testing for reliability reasons at the end of the treaty.[87] This decision was then promulgated in Presidential Directive/NSC-38 on 20 May.[88] As noted previously, Carter also decided that nuclear weapons experiments at minimal-yield levels (a few pounds or somewhat higher) should be permitted under the CTBT in addition to experiments in laser fusion and other related areas for civil energy purposes. He directed that the precise nature and yields of such experiments be detailed in a CTBT Safeguards Plan by the inter-agency Special Coordination Committee and forwarded for his review by 30 June. However, the Departments of Defense and Energy as well as the JCS were far from happy with this.[89] The JCS even sent their own written response to the President, forwarded by the Secretary of Defense Harold Brown, which made clear their strong opposition to his decision. For the JCS the magnitude of the risks and the potential consequences compelled them to conclude that the US negotiating position, as articulated in NSC-38, could result in a treaty that would adversely affect US national security interests.[90] Brown himself noted that the treaty's proposed five-year period instead of a three-year period that had alternatively been proposed reduced the chances of keeping viable nuclear laboratories together – this was his main concern. Los Alamos and Lawrence Livermore laboratory directors had told the JCS that, given a commitment to resume testing, they could maintain relevant laboratory effectiveness for three years, but probably not for five. Brown therefore urged that these two issues be re-examined as part of the development of the safeguards programme requested by Carter for June 30.[91]

London quickly became aware of these retrograde developments.[92] Carter's decision posed some difficult questions for the UK, not least of these concerned what precisely the President had in mind by the phrase 'nuclear weapons experiments at minimal yield levels' and how the new proposals might be best presented to the Russians.[93] On the former, the underlying British concern related to whether these experiments were to be conducted only in laboratories, or at the Nevada Test Site. If the US were intending the latter for stockpile safety and reliability reasons, then the question would likely arise whether the UK would also need to do so. This would pose awkward problems for Ministers who would have assumed that the only thing to have changed with Carter's announcement was the treaty's duration, not its comprehensiveness.[94] And to make matters worse, there were signs that differences were now emerging from within the US Administration on the size of the experiments to be conducted.[95] This could well lead to further delays in the treaty negotiations. Dr David Owen still hoped, however, that the treaty would continue automatically after the five-year period. Macklen, on the other hand, was concerned that the US was arguing itself back to the original Soviet proposal of a three-year treaty, a position that the UK had rejected.[96] Coping with the ins and outs of the wayward Washington policy process made life difficult for Ministers and officials alike. At times the 'special relationship' seemed no more than a ticket to the front row of the theatre that is the US interagency process. Whether it was a Shakespearian tragedy, or Commedia dell'arte is not clear.[97]

As a result of these disputatious representations by the agencies and labs, the President held a further meeting on 13 June to review these issues with the laboratory directors and the Secretaries of Energy and Defense.[98] Subsequently, a Special Coordination Committee meeting, chaired by Brzezinski, considered the next steps on treaty duration, permitted experiments and verification in light of recent meetings and deliberations by the President. Carter was now of the view in light of the 13 June meeting that kiloton-level testing should still remain out of the equation, but he had more flexibility in considering a shorter duration treaty with very low-level permitted experiments.[99] This meeting, which included representatives drawn from all the key inter-agency players, agreed to make changes to the current version of PD/NSC–38; namely, the treaty duration would be three years instead of five years, a deletion from the intention to resume testing clause of the restriction to reliability and safety purposes, and raising the level of permitted experiments to less than one hundred pounds instead of a 'few pounds or somewhat higher'. Yet even this did not quite assuage the anxieties of the Los Alamos and Lawrence Livermore laboratories. In response, they argued that the shorter period for the CTBT only alleviated some of the national security concerns. It is conceivable that this was a case of protesting too much, and that basically the labs would prefer uncon-strained testing of the sort that they were used to. Nonetheless, they insisted, a noticeable risk remained if testing of the performance of boosted primaries was foreclosed, since it was now thought that there was a low probability that some deficiency could occur in critical weapons in the US stockpile. It would seem essential in their view that the Administration should look to articulate the compensating benefits of a three-year moratorium instead.[100] A further meet-ing of the Special Coordination Committee was held on 6 July in an attempt to thrash out a common US position that all the agencies could agree upon; but the JCS remained outside an emerging consensus that a three-year treaty was acceptable without undue risk to US security.[101] Dr Brzezinski suggested that the Committee should now report to the President that all agencies were prepared to sign on for the three-year approach and that the JCS was standing by its concern, but were also more satisfied with the three-year than five-year proposal. All this echoes the similar divisive debates in the US during the late 1950s and early 1960s when there was a strong anti-CTBT lobby in the US. It is also odd that the JCS, rather than the scientists directly concerned with warhead design should be the ones most animated by the prospect of even a limited test ban treaty, but then again, the US JCS could always be relied upon to take a hard and inflexible line.[102] Perhaps the main underlying concern here was the military need for wholly reliable weapon systems, which would ensure that their operational plans could be fulfilled if nuclear war were ever to break out. In a counter-force strategy, this sort of thing no doubt mattered.

This new outcome confronted the British with some pressing problems as they had repeatedly emphasised to the US their strong view that five years was the minimum term which should be negotiated.[103] The first of these problems was what line should the UK adopt in a plenary statement in the tripartite negotiations in support of the US position. Second, what further steps should be taken with the US on treaty text on permitted military and civil nuclear experiments and on the verification implications for a five-year treaty? Britain also wanted to ensure that safety tests, of the sort that were conducted during the 1958–1961 moratorium, should be excluded from any test ban.[104] Third, what should be reported to Ministers and finally what line should the Prime Minister take when he met his Indian counterpart? Discussions with the Americans had shown two aspects of the President's decision uppermost in their minds. First, any arrangements that followed a five-year treaty would need to be ratified by the Senate and this would be the case whether those further arrange-ments took the form of an exact copy of the present treaty, an amendment to the present treaty, or a completely new treaty. Second, the US would resume testing at the end of the five-year period unless to do so had in the meantime been shown to be unnecessary, and this approach to the possible continuation of the treaty would be made public. It was plain to the UK, however, that the US had not thought through its position on a replacement treaty, and this perhaps was somewhere the UK could help by offering to draft text with the US to help clarify the problem. Much more problematic were the various indications from the US that despite the President's decision, arguments still raged within the agencies. Some British officials were of the view that the decision had led to the worst possible outcome as it would neither address the stockpile safety and reliability problem nor address the non-proliferation imperatives either. GEN 63 now thought that the UK should use the phrase 'a treaty of five years' duration' in any discussion on the immediate point of duration and should also seek to soften any treaty text on a replacement treaty. So far as possible, the UK would aim for a form of words that did not imply that there would necessarily be some arrangement to follow the expiry of the treaty. However, it would leave open the possibility that there *might* be a continuation of the treaty in its present form. Officials also thought that it would be desirable to use the term 'conference of states parties' – something that would be more likely to appeal to NNWS as it would involve them in the process – for the review that would take place at the end of the five-year period. Such language could read thus: 'During the five years of the treaty there should be a conference of parties to the treaty to determine whether some further arrangements should follow after the 5 year period'. Clever drafting can certainly help, and can even paper over the cracks when negotiating a treaty or agreement, but equally even the flexibility and richness of the English language cannot stretch to bridge the unbridgeable. Things were now at the unbridgeable stage, if indeed they had not already reached that point.

Until now the UK had refrained from intervening in the internecine squabbles in Washington. The Prime Minister was scheduled to see President Carter in Bonn on 16 and 17 July and this would be the perfect opportunity to find out what the US position now was, or likely to be. That assumed of course that the President actually knew what this was given the fractious nature of the interagency process, but evidently a CTBT of indefinite duration was totally out of the question and instead only a shadow treaty was likely to be achievable. A treaty that had no non-proliferation value, such as what was emerging from Washington, would not appeal to the Ministers of a Labour government for whom arms control and disarmament were important policy goals. Once again British officials believed that it was imperative that the US consult the UK before putting a position to the Russians. Callaghan asked Carter in Bonn during the Economic Summit on 16–17 July 1978 for US views on permitted experiments – and pressed for early consultations if the US position was now for a treaty of only three-year duration. Callaghan stressed that this would not appeal to NNWS and strongly emphasised the UK concern that there should be no announced intention of resuming testing at the end of the CTBT's term.[105] He had told Carter that he was afraid that NNWS would be deterred from supporting a treaty if the President found it necessary to make a statement that the US would be likely to resume testing after the expiry of the treaty. Unfortunately, even lobbying at the highest level in Washington did not lead to the sort of outcomes that the British had hoped to see. The special relationship, even the nuclear one, evidently had its limits. As officials recognised, the UK had little choice but to accept these changes to US policy.[106]

Permitted experiments

MOD promised GEN 63 in September 1978 a note setting out the case for the UK to support the US proposal for 'permitted experiments' for defence purposes during the currency of a test ban treaty.[107] The US had proposed during September in the tripartite negotiations that such experiments should be specifically catered for in a CTBT.[108] In essence, the US suggested that laboratory experiments with nuclear yields not above 100 lb should be permitted.[109] It seemed therefore that the inter-agency argument on this topic in Washington appeared to be over, and it seemed too that the President would decide that experiments with a yield up to 100 lb should continue under a CTBT.[110] As for UK requirements for such experiments, GEN 63 recorded that there was certainly a strong case for them. The military arguments included the need to maintain the expertise of the scientific staff involved with nuclear weapons by giving them work in a field relevant to weapons. Safety considerations might also provide a case for permitted experiments. However, exactly where such work should be conducted posed problems for the UK. Ideally, some of these should be done at AWRE, whilst others should be performed at the Nevada Test Site. Nuclear experiments with very low yields already took place at Aldermaston, and to conduct experiments with yields up to 100 lb would only be an extension of the then-current practice.[111] As for the public presentation of such things, officials recognised that any statement should make clear that experiments with minimal yields such as 100 lb would not detract from the comprehensive nature of the treaty – echoes of the Maralinga Experimental Programme in the late 1950s and early 1960s where minor trials involving very small quantities of fissile material had taken place in order to address warhead safety issues amongst other things.[112] Officials thought that it would be better not to deal with this directly in the treaty, but rather the whole question of permitted experiments should be discussed informally with the Russians. These questions – permitted experiments, yield and where the UK might conduct such things, public presentation and handling with the Russians – would all need to be addressed by Ministers and British positions for the negotiations agreed. Under a CTBT the MOD would have to solve the problem of maintaining the reliability and safety of the nuclear weapons in the operational stockpile namely Polaris and the WE177A, B and C weapons. The use of the term 'operational' is significant as clearly warhead spares were part of the overall stockpile, but as such could be more readily monitored or broken down as required. Weapons built in the late 1960s and early 1970s used exotic materials, some of which were chemically active – the interactions with these on other components could lead to corrosion and other undesirable chemical reactions elsewhere in the warhead, both of which could impact reliability and safety as these changes could cause the finely engineered weapon tolerances to drift off specifications. As already noted, the weapons surveillance programme was designed to counter this and entailed weapons being withdrawn from service and broken down with all component parts – nuclear and non-nuclear – subject to close forensic inspection and testing to look for changes that could impact weapon integrity. In the great majority of cases, corrective action could be taken confidently on the basis of knowledge already available to the designers. Unusually some action might require a nuclear test, but it was much more likely that any modifications would be included in a test device that had been intended primarily for other purposes. The success of the surveillance programme was critically dependent upon the competence of the weapons designers (and engineers) in the UK and it was these individuals who were also responsible for the viability of current weapons and for designing new ones. Reliance on the US for advice would be important too, given its much larger range and breadth of experience with comparable issues. Responsibility for the very large number of non-nuclear components in the WE177 weapons lay with Hunting Engineering, and none of their experts were involved in full tests of nuclear devices.[113] A significant part of weapons reliability and safety work therefore resided outside AWRE though with a company that had previously been directly involved in the development and manufacture of the WE177 weapon system.

Aldermaston scientists could see that a number of permitted experiments would be required under a test ban to help ensure reliability: use of inertial confinement fusion, which over the years would play a role and could become an appreciable fraction of the 100 lb limit that the US was seeking; testing of implosion systems without radioactive materials, or at least using only very small quantities, which could produce a very small yield to assist in the diagnosis of implosion efficiencies. MOD was anxious to avoid forgoing the right to carry out safety experiments on the grounds that there might be a risk of a resultant small nuclear yield. If the US were to build special facilities for these sorts of activities – and there was some indication that one might be constructed at Nevada, then the UK would hope for very close collaboration with the US and to take advantage of any new facilities.

The US opts for a three-year treaty: September 1978

President Carter eventually decided to opt for a three-year treaty at the end of September 1978. He simply could not ignore the level of sustained and bitter opposition to a five-year ban coming from the JCS, and were such a treaty ever to make it to the Senate floor for advice and consent on ratification, such opposition would become public and very likely scupper any chance of Senate support. However, Carter did not take any new decisions on the resumption of testing after the treaty's expiry, the terms of reference of the Review Conference and smaller nuclear experiments that might be permitted during the treaty. The Americans planned to inform the Russians of this new position when the tripartite negotiations resumed in Geneva on 28 September 1978. Meanwhile, the UK recognised the reality that it would have to acquiesce in Carter's decision even though it would reduce the non-proliferation value of the treaty, but this was not the most important point for the UK as the other taxing questions still had to be addressed. London certainly expected that the US would consult its British allies before coming to a final view on them.[114] As we have seen, there could be no guarantee that the US would do this. Callaghan told Carter on 14 September that:

> I believe you will also be considering the question of 'permitted experiments'. I think that the yield limit of 100 lb which was mentioned to Gromyko is low enough for us to be able to claim that the test ban really is comprehensive. But a higher limit would make the treaty appear to be a threshold treaty, which you and I have decided against, and this, I know, would be badly received by countries like India.[115]

Uncertainty in the US position appeared to be never-ending. The President's response to the Prime Minister on 26 September had stated that his preference was that when the negotiations resumed, the US and UK should negotiate on the assumption that the duration of the agreement will be three years.[116] He also noted that he had not yet completed his review of all CTBT issues. For UK officials the US statement that it favoured the lower threshold could not be taken to mean 100 lb. General Ed Giller of the JCS had indicated to an unnamed UK official that, in the view of the Joint Chiefs there was no final decision to stay below this level and that they would fight for something much higher.[117] London felt that it would not be wise to probe too deeply on this, but it was important to find out whether the Americans had advanced their thinking on how to handle the 'permitted experiments' issue in the Geneva negotiations. UK preferences remained squarely on the need for a very informal written understanding with the Russians that would receive minimum publicity. And this seemed to be the US view too, but ACDA doubted that this could be kept low profile as it would have to emerge in any subsequent Congressional hearings on treaty ratification.[118] General Giller himself subsequently made clear to Denis Fakley (MOD, but also acting as deputy head of UK test ban treaty delegation) in Geneva that there was no longer any active debate about the 100 lb limit for a three-year treaty.[119]

By early November 1978 the US still had not resolved its own position on what would happen after the end of the three-year duration and it appeared that the issue was still some way from resolution.[120] Similarly the US proposals on permitted experiments also still required clarification in British eyes. All of this meant a delay in the tripartite negotiations – allied to the departure of Paul Warnke, the head of the US delegation, which would make it impossible to complete a treaty text before April 1979. The permitted experiments issue was the sole item on GEN 63's agenda when it convened on 10 November 1978. Officials met to review the paper prepared for Ministers that set out policy considerations and options for the UK. Several significant amendments were made and agreed.[121] These concerned the need to avoid compressing the background to the issue – the original US working paper from December 1977; the need to highlight that the UK position was based on what the Prime Minister had told President Cater on the subject back in September 1978; it would be impossible to prohibit or verify 100 lb experiments; a cooperative programme with the US would be important to the UK – coordinated programmes with the US would be good, especially if this entailed use of the Nevada Test Site and finally, that Inertial Confinement Fusion R&D for civil uses should not be constrained. The Cabinet Office revised the paper and a final version was completed on 23 November for submission to Sir John Hunt, the Cabinet Secretary for onward passage to the PM, Defence and Foreign Secretaries.[122]

Permitted experiments were of crucial importance for the UK, especially when conducted jointly with the US; indeed the US wanted to see UK experiments conducted as their results would serve as a sort of a peer review of their own internal studies on stockpile reliability issues – an added incentive for the UK to have its own programme at AWRE, and another example where the US saw value of a UK contribution to its own weapons programme. Any public statement on the need for permitted experiments would inevitably attract criticism from the NNWS even though such low-level experimental work could not be used to advance weapons development work. Officials concluded that there should be no dilution in UK support for the concept of a CTBT. Provided no reference to 'permitted experiments' was made in the treaty, UK participation in a programme of such experiments would not be inconsistent with British basic assumptions on the treaty's scope. Therefore, the options for the UK were clear. Ministers were invited to endorse these: it should support in principle the US position and keep open the option to conduct small nuclear experiments either alone or in cooperation with the US. Three action points stemmed from this, all of which involved further discussions with the Americans. As already noted, the UK inevitably had to spend much time in consultation with the US, and here was yet another example. One might suppose that if the US policy process were less dysfunctional, then there might have been less need for so many bilateral discussions. The key questions this time around included the conditions under which 'permitted experiments' should be conducted; how the proposals should be handled in the tripartite negotiations; and what line the UK and US should take internationally with the NNWS and in public.

Throughout the drafting of the GEN 63 paper, FCO officials had reserved the Foreign Secretary's position and in their advice to Dr Owen, the Arms Control and Disarmament Department suggested that whilst there need be no objection to further discussion on 'permitted experiments' with the Americans, the decision on whether the UK should conduct such experiments should be kept open for the time being.[123] FCO officials reckoned that a major effort would be needed with the NNWS who would take the continuing nuclear experiments as a reason not to join such a treaty. But it was hard to challenge the GEN 63 paper's conclusion that since the US, and almost certainly the USSR, would conduct such experiments, abstention by the UK would not persuade more NNWS to sign the treaty; and that therefore the UK should discuss aspects of this issue further with the Americans. Owen saw 'no need to consult for permitted experiments. Certainly not on the basis of this paper (i.e. GEN 63). If the Americans decide that they will, then we will if we want. But there is no need to press for it. If they can live without, so much the better'.[124] Exactly what officials made of this is unclear.

Although President Carter had confirmed the figure of 'up to 100 lb' as the yield for permitted experiments, the UK had still not been informed of any precise formulation of what the Americans envisaged being covered by the term 'permitted experiments', nor despite repeated enquiries had they been able to offer any clear ideas on how they would propose that the question should be handled in the tripartite negotiations with the Russians or publicly.[125] The UK pushed the US hard on the need to establish precisely what it meant by permitted experiments and to work out the best approach to be adopted in the Geneva negotiations. For the UK at least the purposes of permitted experiments were clear; they related to nuclear weapons stockpile maintenance. As noted earlier, the same arguments and concerns remained in the MOD and AWRE and these were aired repeatedly in intra-Whitehall meetings and in advice to Ministers. So, for example, we hear again that modern weapon designs involved the use of exotic materials and fine engineering tolerances. Great care had to be taken in both the design and stockpile maintenance to avoid the risk of corrosion or instability developing whilst the weapon was in the stockpile. The most effective way for dealing with this would be to begin a research programme that exercised as many as possible of the skills required short of the tests that would be prohibited by a test ban treaty. Inevitably close cooperation with the US would be crucial here since American practical experience of solving stockpile problems was considerably greater than the UK's. A yield limit of 100 lb limited the range of work that could be conducted and such experiments could not lead to the development of new warhead designs. However, officials recognised that future breakthroughs might make such limits less of a constraint in future. The sort of work that could be carried out under these constraints included: the process involved in inertial confinement fusion, which could lead to a greater understanding of weapon physics; the simulation of implosion engines that can give very small nuclear yields; shock tests of fissile materials to help in producing computer simulations for warhead designs; and, safety experiments where a zero yield is expected, but where a yield of some tens of pounds could occur.

Although British officials believed that it would be consistent with the UK approach to the concept of a comprehensive test ban to decline to associate itself with any understanding between the US and USSR on permitted nuclear experiments and to refrain from conducting them, such a position would be unlikely to convince the NNWS to join the Treaty. Moreover, abstinence would only undermine the UK's own weapons programme, so for these reasons, the only options open for the UK were to support in principle US proposals for permitted nuclear experiments and to conduct such experiments in the UK. If Ministers were to accept this conclusion, then the UK would need to discuss three key questions with the US: Under what conditions should 'permitted experiments' be conducted? How should this topic be handled in the tripartite negotiations? And what line should the UK take both internationally with the NNWS and in public?[126] That the UK was going over the same ground again in late autumn 1978 testifies to the continuing failure of the Americans to commit to a clear position – Britain's questions and concerns were going unanswered.

These points were agreed by Ministers in November 1978.[127] The Defence Secretary emphasised to his colleagues the importance which he attached to the contribution that permitted experiments would make towards MOD staff's ability to retain an adequate capability to ensure that the UK's nuclear weapons stockpile remained safe and that any degradation in its operational reliability was kept to a minimum. Furthermore, Mulley also highlighted the importance that under a CTBT the UK continued to draw as much benefit as possible from a UK/US exchange of nuclear weapons technology. The extent to which the UK succeeded in this was likely, in Mulley's view, to be largely influenced by the degree to which the UK participated in a joint programme of 'permitted experiments'.[128] David Owen, in contrast, thought that the UK should not discuss 'permitted experiments' with the US. He was concerned at the negative effect a programme of such experiments might have on the attitude of non-nuclear weapon states to the Treaty. However, he did agree that both MOD and FCO officials should discuss this sensitive issue with the US, but he did not believe that the UK should make any assumptions at that stage about whether the UK should also conduct a programme of such experiments and recommended that decisions on this point should be deferred until the UK had a more detailed definition of the scope and purpose of any programme of experiments.[129] Consequently, FCO officials were tasked to arrange yet another bilateral meeting with the US in Washington in the week beginning 15 January 1979 to discuss these and other CTBT matters. Briefs setting out the points to make and lines to take were prepared accordingly by the MOD and FCO.[130] The three key questions that the UK wished to have answers from the Americans were:

- What they could say about progress in the technical studies upon which they were currently engaged into the requirements for 'permitted experiments' under a CTBT?
- How far did these studies embrace both civil and military requirements?
- Had the US reached views on the conditions and locations for such experiments?[131]

MOD officials knew that ACDA had been trying to place strict constraints on what the US could do under a permitted experiments regime: experiments had to be fully contained, were to be carried out above ground, were to be conducted in re-usable facilities, were not carried out at recognised test sites and must not be used for weapons design purposes. ACDA's conditions were, unsurprisingly, opposed by the JCS. The only place in the US where 100 lb explosions could take place was at the Nevada Test Site (NTS) in either a borehole or tunnel.[132] UK interest in the answers to these questions was without commitment to any British programme of permitted experiments.[133] As for the tripartite negotiations, the subject could not be avoided any longer. The Russians too were still pressing for clarification on the December 1977 US statement about a need for 'mutually acceptable understandings regarding distinctions between prohibited explosions and permitted experiments'. For the British, any understanding along these lines should be as informal as possible. Furthermore, the subject should not be mentioned in the Treaty. However, the Russians had shown no interest in any understanding and if they expressly dissented from a US proposition that such experiments were permitted, this would weaken the legal case for conducting them. That being the case the UK mused that perhaps it might be best simply to opt for a simple unilateral statement in the negotiations that could be understood to apply to permitted experiments and hope that it would pass without adverse Russian comment. From a British perspective, there were three other relevant points here. Officials thought it reasonable to take the view that small experiments would fall outside the scope of the test ban just as they did under the PTBT. There was an underlying, though unstated, assumption in the negotiations that very small experiments were of little concern – an assumption supported by the practical irrelevance of such experiments to the treaty verification procedures being negotiated. And finally at no stage had anyone contemplated in the negotiations that a future CTBT would for example halt such activities as the Joint European Torus research project at Culham on controlled fusion.

The permitted experiments issue was of course closely linked to the language that might be set out on the decisions to be taken by a review conference after the passing of the initial period of three years – although the UK favoured five years, it was now prepared to negotiate for three, which, as we have seen, was strongly favoured by the US. After that, it could be allowed to lapse, be renewed or be modified in some way. All options should be left open in the US view of things. This led to one of the key unresolved issues – both domestically in the US and in Geneva – the language to describe the function of the proposed Review Conference at the end of the three-year period.[134] For the UK, for both non-proliferation and negotiating reasons, it would be much more preferable to have text that provided for all options, but pointing towards extension. However, there was some nervousness about this whole issue in Washington.[135] The US however had still not excluded the possibility that there might be something written into the treaty on permitted experiments. Since the US would definitely conduct such experiments, it was important to place something on the matter on the negotiating record, but the US was aware that the UK preference was for an uncontroversial unilateral statement. Dr Herbert York, the new head of the US CTBT delegation, opined that this could well end up being the US preference too. Bob Press, Cabinet Office and a player in the 1958–1962 test ban negotiations, suggested however that it would be best to avoid extending the scope of permitted experiments too widely. In his view it would much more desirable to stay within the realm of the strongest justification for them which was, while nuclear weapons were retained, to preserve such technical expertise as was necessary to ensure the best chance of maintaining both safety and reliability.[136] No one, least of all in Washington, seems to have worried that a three-year treaty was very likely to have no real impact as an arms control measure.

By March 1979 the US still had nothing to say on the vexatious problems of permitted experiments. This persistent inability of the Americans to get their act together was proving to be one of the biggest obstacles to progress. It is always easier to obstruct than to make progress in any negotiation and indecision such as 'we have no instructions' as well as wilful obstructionism, poses an equal barrier to agreement. John Edmonds, Head of the UK delegation, pressed Dr York at UK–US delegation meetings in Geneva for any updates and on both occasions he was so pressed, York had still nothing to report.[137] Following his return from consultations in Washington, York told Edmonds that he would do his best to sort out the US position on permitted experiments during the coming recess in the negotiations. He emphasised that the problem was an extremely difficult one on which the US SCC would have to meet; and when it did it might well be split, so that the issue would have to go to the President for a determination. As we have seen even a Presidential determination on this issue did not resolve things. According to York the first question concerned the stockpile; the weapons laboratories also wanted to conduct small tests to try out 'minor new features' such as fitting particular weapons to particular delivery systems and new ideas on safety. He realised that the UK and USSR saw no need for any 'safeguards' programme, including permitted experiments to be made public. York feared that there was, however, no possibility of stopping the fact of these experiments from becoming public knowledge in the US. His own view was that the US would have to explain its programmes to the UK, then the Russian and the other governments in NATO before it became public.[138]

We can also see that by the end of April 1979 there still had been no decision on any UK programme of permitted experiments as officials began to prepare briefs for an incoming government following the calling of a General Election for 3 May 1979.[139] Notwithstanding York's hopes, the related issue of permitted experiments and stockpile reliability were still amongst the outstanding issues in the negotiations in the middle of July 1979.[140] And by the close of the year we still see UK officials noting the need to press the US for decisions on how to handle a requirement, which the UK shared, for small nuclear experiments to continue under a test ban. Both FCO and MOD officials felt that a simple unilateral statement in the negotiations, which would be understood to cover 'permitted experiments' and would be designed to pass without contradiction by the Russians still seemed to be the best approach, but the new government of Margaret Thatcher had not yet confirmed this.[141]

A new broom: Margaret Thatcher and the Conservatives take the reins

In keeping with tradition, civil servants in the Cabinet Office, FCO and MOD had prepared papers on policy issues and choices facing incoming new Ministers when they assumed office after the 3 May 1979 General Election. Margaret Thatcher, the new Prime Minister, was quickly briefed on the state of play on the CTBT and the pressing issues of permitted experiments. Within five days of assuming office, the Prime Minister had made clear that in her view the proposed 100 lb limit on small nuclear experiments was too low. Moreover, the UK should not give way on the question of nuclear experiments and should work to exclude from the Treaty much bigger tests than so far envisaged since the Russians had a capability to decouple their test explosions (conducting tests in large cavities, thereby reducing the amount of energy transferred to the surrounding rock, which would reduce the seismic signature of the event). Thatcher also remarked that the three-kiloton threshold for tests below which testing could be resumed was too low – it should be 10 kilotons.[142] She repeated the same point in her meeting with US Secretary of State Cyrus Vance on 23 May 1979.[143] It is not clear on what technical basis the Prime Minister was using for such judgements, but considerations of Polaris replacement and the need for the UK to continue testing may have been a major part of this as this issue was also one of the pressing nuclear issues requiring attention from the new Prime Minister. In her meeting with Cyrus Vance, she had stated that she had private scientific advice unknown to the Foreign Secretary.[144] Just what the source for this private advice was remains unclear. Thatcher also said at this meeting that she was concerned that the Soviet Union would make elaborate preparations towards the end of the treaty period for a comprehensive testing programme that could begin as soon as the treaty expired.[145] This was an allusion to what had happened in September 1961 when the Soviet Union broke the testing moratorium – and was the sort of experience that had lived long in the memory of at least one senior and influential nuclear official working in the Cabinet Office.[146] Indeed UK policy on the role of a moratorium in a new CTBT negotiation and in the period prior to entry into force had been directly shaped by this bad experience.[147] In short, the Prime Minister was 'unhappy and concerned' about the CTBT.[148] In fact, what the Prime Minister actually wrote on the test ban negotiations brief was that she was '*very* unhappy and suspicious about this'.[149] She had concluded that the UK should not sign a CTBT unless it could be assured that it could test its stockpile and maintain a national competence as well as the Soviet Union; and the Prime Minister also made this abundantly clear to President Carter's Scientific Advisor, Dr Frank Press, in mid-June 1979.[150]

MOD subsequently told No 10 that it was the considered technical judgement of both the US and the UK that it would certainly be possible to maintain the reliability and safety of the existing Western nuclear stockpiles in the long term if underground testing up to a limit of five kilotons were permitted, and it was possible that even three kilotons might be sufficient.[151] For the MOD permitted experiments up to 100 lb would be enough to maintain technical assessment capabilities for stockpile reliability and to maintain expertise necessary to permit a resumption of warhead development and production. There was little advantage in extending this to 500 lb, or even several tons. Moreover, MOD did not seek a 10-kiloton threshold as this could work to the advantage of the Russians since it could enable greater design options and the possible masking of the yield through evasion techniques such as decoupling.

The CTBT negotiations remained in stalemate in the middle of 1979. One of the main problems was the UK's refusal to accept more than one national seismic station on its territory as we saw in Chapter 2.[152] But there were other problems of greater import. There was no solution in sight on the permitted experiments problem, or on what would happen after the end of the three-year ban. It seemed to the UK that the Russians were in no hurry as they were becoming concerned that the US would be unable to agree on a test ban and with an impending Presidential election in November 1980, the prospects for a test ban treaty were growing slim in more general terms.[153]

Critical advice on the need for testing in the context of stockpile reliability and future warhead designs in a test ban was provided to Ministers by an advisory committee chaired by Lord Penney in 1980 and was a key part of a Cabinet Office-led review of UK testing policy in January 1981 whose aim was to provide policy recommendations for Ministers in light of the state of play on testing issues.[154] Penney's panel noted that the mechanical properties of the materials used in a nuclear warhead were likely to change with age, but it was virtually certain that any faults arising within a period of three years could be remedied by AWRE without recourse to nuclear weapon testing. The warhead surveillance programme was largely designed with this objective in mind.[155] From about 1961 to 1976 some 20–30 in-service faults or faults had been detected with about ten of these having some safety significance whilst the remainder affected reliability.[156] There was no reason to suppose that this position would change in the subsequent three years provided the component materials continued to be available to reproduce the original designs and thus to reconstruct the weapon to its original specification. However, there were components in a warhead whose performance depended critically on both their chemical stabilities and their precise spatial relationship. It was conceivable that circumstances could arise, through for example, alterations in the availability or purity of the materials used when AWRE would not have its present degree of confidence in its ability to reproduce exactly the original specifications. In such circumstances, Penney's panel accepted that AWRE would then have to advise Ministers that a nuclear test would be desirable. Overall, the Panel concluded that nuclear tests were not necessary to maintain capabilities in material control, fabrication, safety operations for radioactive or explosive processes, inspection and measurement of material properties. However, nuclear tests would be necessary to justify the introduction of a new design into the stockpile, if at least as sophisticated as those designs that were already in service with the Royal Navy and RAF. Furthermore, it could not be assumed that a US-tested design would be made available to the UK; and even if it were, the Panel thought that a UK copy would not be acceptable for the stockpile without prior testing. This comment was pivotal and reinforced existing AWRE and MOD assessments of what would be required for a future Trident warhead. By this time the threat posed by a time-bound or indefinite test ban treaty had long since ceased to be a threat to the UK's nuclear weapons programme.

Conclusion

It was clear at the time, and indeed more or less from the outset of the tripartite negotiations, to UK Ministers and officials that what was needed for stockpile safety and reliability question undermined the quest for a CTBT. No Minister, Labour or Conservative, could disregard technical advice on the fundamental questions of nuclear weapon stockpile reliability, safety and maintenance even if they might be doubtful about the real strength of the case being put to them by the nuclear weapons scientists. Owen and Zuckerman certainly felt that there was some sort of conspiracy by the weapons laboratories on both sides of the Atlantic to thwart their test ban aspirations.[157] However, in retrospect, it does not look as if things were as clear-cut or as simple as that. Five points seem to be germane to an overview of what happened here. First, there had always been a strong animus against a test ban treaty in the US – both within the laboratories and elsewhere in the defence communities as we saw in the late 1950s and early 1960s and one again in the late 1970s. This had not disappeared and would not have required any encouragement from AWRE or MOD scientists. Second, Lord Zuckerman had not endeared himself to AWRE and MOD when he was Chief Scientific Adviser to the government in the 1960s and seemed then to be determined to reduce AWRE's role and the money spent on UK nuclear weapons on the grounds that he saw limited utility for them in any credible British defence policy. Although he had dealt directly with nuclear weapons issues as MOD's Chief Scientific Adviser and acted as the UK principal in the annual stocktaking meetings with the US under the 1958 Mutual Defence Agreement, Zuckerman was not best placed to comment in the late 1970s on the technicalities of the stockpile reliability and safety issues as he had long since ceased to have any direct responsibilities for them. Whilst he could often find the right question to ask, he was not always right in some of his assumptions and tended to be dismissive of the scientists at AWRE. He never seems to have grasped the subtlety of the technical argument that it was the confidence in the judgement of the experts who certified the stockpile that arose from having their predictions and assessments validated by periodic full-scale tests that was at the heart of the argument over reliability and safety.

Third, whilst there is no evidence in any of the open files at The National Archives that there was any sort of deliberate conspiracy between AWRE and MOD and the US labs, one might not expect in any case to see people write such things explicitly in papers put on file. Not all conversations are minuted. As the key interdepartmental body that chaired and coordinated British test ban policy, GEN 63 was composed of both senior MOD and Foreign Office officials; it was chaired by the Cabinet Office, and Sir Clive Rose himself was on secondment from the FCO. The record of GEN 63's deliberations clearly shows that all decisions on policy and recommendations for Ministers were carefully debated and agreed, often at great length with papers undergoing repeated redrafting and appearing in numerous new iterations. The very nature of the British system was to produce interdepartmental agreement on policy recommendations to Ministers laying out clearly all the pros and cons. We can see some signs of growing disagreement, – particularly in 1980,[158] but it is not clear that this was necessarily because of diametrically opposed arguments between the FCO and MOD. Even before the negotiations formally began in the summer of 1977, the UK had considered the reliability and safety arguments *before* entering into the test ban treaty talks in Geneva, and at that point, officials had wanted to discuss the matter with the US. So in that respect, this was not a new issue that suddenly appeared out of the blue, or had been provoked by the AWRE or the MOD by waking up the Americans. Usual inter-agency disarray and disagreement was much more of an explanation for the initial delay in Washington trying to sort itself out on this problem. In fact, it could even be argued that the US never fully sorted itself out on these issues, and this in part explains why the treaty negotiations failed.

Fourth, the UK self-imposed testing moratorium was a result of a policy decision in the mid-1960s that there would be no strategic successor to Polaris; and since the new generation of tactical weapons – the WE177s – were entering service between 1966 and 1977, there were no new designs that needed to be tested or validated.[159] Moreover, the UK had withdrawn older weapons (Red Beard, Yellow Sun Mark II and Blue Steel),[160] so issues of reliability and safety did not arise. Had a major reliability or safety issue arisen within the stockpile a test could have been conducted. Once full nuclear testing was resumed in connection with the Polaris Improvement Programme in 1974, the UK quickly realised what it had lost as a result of its self-imposed restraints.[161]

Fifth, UK nuclear warheads had been designed with the assumption that full testing would be available. Apart from the early generation of nuclear weapons in the UK stockpile, which was only intended to have a relatively short shelf life of about five years or so, later designs for the Polaris warhead (ET317) and the WE177 were designed to last very much longer. Combined with budgetary and domestic political pressures on UK testing, these factors meant that when the UK did plan and conduct a test, Aldermaston tried to gain the maximum amount of scientific knowledge from each test. This helped to ensure that the UK was probably better able to sustain its small programme with fewer tests whereas the US labs appeared to have become used to an almost limitless number of tests and did not seem to have to worry too much about the cost. In the period of the UK self-imposed moratorium from 1965 to 1974, the US conducted 389 tests.[162] For all that, the stockpile reliability and safety issue very clearly put paid to any chances of a comprehensive test ban treaty being agreed in the tripartite negotiations between 1977 and 1980. It was hardly a surprise then when the Reagan Administration after taking office in January 1981 conducted a review of US testing policy and concluded that a CTBT was not in US national security interests. This new policy was announced to the Committee on Disarmament in February 1982. The world would have to wait another 15 years before a truly Comprehensive Nuclear Test Ban Treaty was adopted by the UN General Assembly on 10 September 1996 following almost three years of negotiations in the Conference on Disarmament from February 1994 until

August 1996.[163] By that stage, the nuclear weapons laboratories had clearly found ways to assure themselves sufficiently of stockpile reliability and safety and were funded accordingly. Aldermaston embarked on a Warhead Assurance Programme designed to 'ensure the safety, effectiveness and durability of the UK nuclear warhead stockpile'.[164] Changed days indeed.

Notes

1 See John R. Walker, *British Nuclear Weapons and the Test Ban: Britain: The United States, Weapons Policies and Nuclear Testing: Tension and Contradictions*, Farnham, Ashgate, 2010.
2 TNA PREM 19/212, R.L.L Facer, MOD to B.G. Cartledge, 10 Downing Street, Nuclear Weapon Stockpile Maintenance, 15 June 1979. For an academic review of the stockpile reliability issue see Steve Fetter, 'Stockpile Confidence Under a Nuclear Test Ban', *International Security*, Vol. 12, No. 3, pp. 132–167, 1988 and 'Correspondence: An Exchange on Stockpile Confidence', John D. Immele and Paul S. Brown, Steve Fetter, *International Security*, Vol. 13, No. 1, pp. 196–215, 1988.
3 TNA FCO 66/875, CTB Negotiations: Implications for Defence Interests, Note by the Ministry of Defence, 1 July 1977.
4 TNA DEFE 23/218, Draft letter from Secretary of Defence to Prime Minister under Cover of a Submission from J.D. Bryars, AUS (Defence Staff) to PS to Secretary of State, Safety and Reliability of the United Kingdom Nuclear Warhead Stockpile under a Comprehensive Test Ban (CTB), 15 March 1978.
5 TNA DEFE 24/1344, Annex B, Comprehensive Test Ban (CTB), Military and Security Consequences, Defence Department, 14 January 1977.
6 TNA AIR 8/2785, J.D. Bryars, AUS (Defence Staff) to PS to Secretary of State, Safety and Reliability of the United Kingdom Nuclear Warhead Stockpile under a Comprehensive Test Ban (CTB), 15 March 1978.
7 TNA CAB 130/952, GEN 63 (77) 1, Nuclear Arms Control, Note by the Secretary, 22 April 1977.
8 TNA CAB 130/952, GEN 63 (77) 1, Nuclear Arms Control, Note by the Secretary, Annex Detailed Aspects of a CTBT, 22 April 1977; TNA CAB 130/952, GEN 63 (77) 1, Nuclear Arms Control, Note by the Secretary, 22 April 1977.
9 TNA CAB 130/952, GEN 63 (77) 6, Position Paper No. 10 for CTB Negotiations: Implications for the United Kingdom Defence Requirements, Note by the Ministry of Defence, 17 June 1977.
10 John R. Walker, 'British Nuclear Weapons Stockpiles by Year: 1953–1977', *RUSI Journal*, Vol. 166, No. 4, pp. 10–20, 2020.
11 TNA FCO 66/898, V.H.B. Macklen, DCA (PN), MOD to Sir Clive Rose, Cabinet Office, CTBT, 23 May 1977.
12 TNA CAB 130/952, GEN 63 (77), 10th Meeting, Item No. 2 Discussions on a Comprehensive Test Ban Treaty, 28 June 1977.
13 TNA CAB 130/952, GEN 63 (77), 11th Meeting, Discussions on a Comprehensive Test Ban Treaty, 7 July 1977.
14 TNA CAB 130/952, GEN 63 (77), A Comprehensive Test Ban Treaty, Note by the Secretary, Advantages and Disadvantages of a Comprehensive Test Ban Treaty, Note by Officials, 15 July 1977.
15 TNA CAB 130/952, GEN 63 (77), 15th Meeting, 8 September 1977.
16 TNA CAB 130/952, GEN 63 (77) 15th Meeting, 19 September 1977.
17 TNA CAB 130/952, GEN 63 (77) 1st Meeting, 29 April 1977.
18 TNA AIR 8/2785, B.R. Norbury, Head of DS 11 to APS/S of S, Comprehensive Test Ban: Warhead Stockpile Safety and Reliability, Background Note, 31 March 1978.

19 TNA AIR 8/2785, J.D. Bryars, AUS (Defence Staff) to PS to Secretary of State, Safety and Reliability of the United Kingdom Nuclear Warhead Stockpile under a Comprehensive Test Ban (CTB), 15 March 1978.

20 Raymond Garthoff, *Détente and Confrontation American-Soviet Relations from Nixon to Reagan*, Washington, DC, The Brookings Institution, 1985, p. 756.

21 TNA CAB 130/1011, GEN 63 (78) 2, Comprehensive Test Ban: "Permitted Experiments", Note by the Secretary, 7 March 1978.

22 TNA AIR 8/2785, John Hunt to Prime Minister, Comprehensive Test Ban: Warhead Stockpile Safety and Reliability, 20 March 1978; TNA DEFE 19/240, D.C. Fakley, D/Ds 6, MOD to DCA(PN), MOD, 13 January 1978; TNA DEFE 19/240, Clive Rose to Sir John Hunt, Comprehensive Test Ban Negotiations, 13 January 1978.

23 TNA FCO 66/1041, FCO telegram No.113 to Washington, 17 January 1978.

24 TNA CAB 130/952, GEN 63 (77), 28th Meeting, Cabinet Official Group on International Aspects of Nuclear Defence Policy, 21 December 1977.

25 TNA CAB 130/1011, GEN 63 (78) 2nd Meeting Cabinet Official Group on International Aspects of Nuclear Defence Policy, 12 January 1978.

26 TNA DEFE 19/240, D.C. Fakley, Director DSc6 to Sir Clive Rose, Cabinet Office, (draft) CTB Negotiations: Definitions, Brief for UK/US Bilateral Discussions, 19 January 1978.

27 An equation of state is a mathematical relationship that describes the state of matter (i.e. gas, liquid or solid) using the material properties of temperature, volume, pressure and internal energy. The equation of state characterises the properties of a state of matter under a given set of physical conditions. National Nuclear Security Administration, 'Office of Research, Development, Test and Evaluation', *Stockpile Stewardship Quarterly*, Vol. 7, No. 4, December 2017, p. 1.

28 TNA DEFE 19/240, D.C. Fakley, Director DSc6 to Sir Clive Rose, Cabinet Office, (draft) CTB Negotiations: Definitions, Brief for UK/US Bilateral Discussions, 19 January 1978.

29 John R. Walker, *British Nuclear Weapons and the Test Ban: Britain, the United States, Weapons Policies and Nuclear Testing: Tension and Contradictions*, Farnham, Ashgate, 2010, p. 94.

30 TNA DEFE 19/240, R.J. Meadway, 10 Downing Street to M.J. Vile, Cabinet Office, 17 January 1978.

31 TNA FCO 66/1041, FCO Telegram No.113 to Washington, 17 January 1978.

32 TNA CAB 130/1011, GEN 63 (78), 3rd Meeting Cabinet Official Group on International Aspects of Nuclear Defence Policy, 20 January 1978.

33 Barry M. Blechman, 'The Comprehensive Test Ban Negotiations Can they be revitalized?', *Arms Control Today*, Vol. 11, No. 6, June 1981.

34 TNA CAB 120/1011, GEN 63 (78) 4th Meeting Cabinet Official Group on International Aspects of Nuclear Defence Policy, 30 January 1978.

35 TNA DEFE 19/240, Telegram British Defence Staff Washington to MOD, 12 January 1978.

36 TNA DEFE 19/241, Comprehensive Test Ban Treaty, Warhead Stockpile Safety and Reliability, amended draft, 10 March 1978.

37 TNA DEFE 19/241, Report on a Visit to Washington to discuss the US Proposals for 'Permitted Nuclear Experiments' in the context of a Comprehensive Test Ban Treaty, R. Press, March 1978.

38 TNA DEFE 19/241, Fred Mulley to Prime Minister, 21 March 1978.

39 TNA DEFE 19/240, Draft CTBT: The Definition Problem, KJ 32/78, 13 February 1978.

40 TNA CAB 130/1011, GEN 63 (78) 2, Comprehensive Test Ban: 'Permitted Experiments', Note by the Secretary, 7 March 1978.

41 TNA DEFE 19/240, A. Reeve, Washington to C. Mallaby, ACDD, Prohibited Explosions and Permitted Nuclear Experiments, 23 January 1978.

42 TNA DEFE 24/1344, Record of a Meeting on a Comprehensive Test Ban (CTB) held in Washington on 14 March 1977.

43 TNA CAB 134/2241, Nuclear Requirements for Defence Committee, Draft British Programme of Underground Nuclear Tests 1965/66, Note by the United Kingdom Atomic Energy Authority, AWRE, 9 December 1964. Until 1973, Responsibility for AWRE had rested with the UKAEA after which it was brought under the MOD.

44 TNA DEFE 19/240, V.H.B. Macklen, DCA (PN) to DUS (P), Permitted Nuclear Experiments – CTB Definitions, 25 January 1978.

45 TNA DEFE 19/240, P.G.E.F. Jones, Deputy Director/Chief Warhead Development, AWRE to Mr V.H.B. Macklen DCA (PN), MOD, UK Stockpile Reliability in a CTBT Context, 1 February 1978.

46 This was simply a programme of continuously refurbishing the stockpile at the lowest meaningful rate to ensure that facilities and expertise were exercised to produce consistent hardware. The complication was that the rate varied from components to component, but the emphasis on consistency was thought to be a vital factor in reliability. AWRE chose a rate of about one-tenth of the stockpile per year for the final weapon assembly, with the sub-components matching where possible, but each being considered on its own merit. Some components were made in batches whereas others with high process control content needed to be made continuously. By this process, the stockpile would be re-lifed over a ten-year period and applicable parts of the weapon being returned from the Royal Navy and Royal Air Force would be available for examination giving confidence in the quality of the rest of the stockpile.

47 TNA DEFE 19/240, Record of a Meeting with Dr Frank Press, Presidential Adviser on Science and Technology Policy, 16 February 1978, 24 February 1978.

48 TNA DEFE 19/240, K. Johnston, AD/DSc 6, MOD to Dr R. Press, Cabinet Office, VISAM 742: Washup Meeting in DCA (PN)'s Office on 22 February 1978, 24 February 1978.

49 TNA CAB 120/1011, GEN 63 (78), 5th Meeting Cabinet Official Group on International Aspects of Nuclear Defence Policy, 23 February 1978. The Report of the Meeting is at TNA CAB 130/1011, GEN 63 (78) 2, Comprehensive Test Ban: 'Permitted Experiments', Note by the Secretary, 7 March 1978.

50 TNA DEFE 23/218, Draft letter from Secretary of Defence to Prime Minister Under Cover of a Submission from J.D. Bryars, AUS (Defence Staff) to PS to Secretary of State, Safety and Reliability of the United Kingdom Nuclear Warhead Stockpile under a Comprehensive Test Ban (CTB), 15 March 1978.

51 TNA DEFE 19/240, C.L.G. Mallaby, Arms Control and Disarmament Department, FCO, Note for the File, CTB Definitions, 23 February 1978.

52 See for example William Epstein, 'Limits on Nuclear Testing: Another View', *Arms Control Today*, Vol. 11, No. 8, October 1981. Epstein argued that US plans to ratify the TTBT and PNET would not have any non-proliferation benefits and would not impress the non-nuclear weapon states. He was right.

53 Lynn R. Sykes, *Silencing the Bomb: One Scientist's Quest to Halt Nuclear Testing*, New York, Columbia University Press, 2017. See chapters 5, 9 and 10 in particular. See also Steven R. Taylor and Peter D. Marshall, 'Spectral Discrimination between Soviet Explosions and Earthquakes Using Short-period Array Data', *Geophysical Journal International*, Vol. 106, No. 1, July 1991.

54 TNA FCO 66/1588, James Harrison, Defence Secretariat 17 to Mrs C.A. Boots, ACDD, FCO, TTBT – Verification and Compliance, MOD Note on TTBT – Verification and Compliance, 3 September 1982.

55 Solly Zuckerman, *Nuclear Illusion and Reality*, New York, The Viking Press, 1982, p. 123.

56 One of the reported reasons for Zuckerman retaining an office in the Cabinet Office was in order to maintain his good personal relations with Admiral Rickover, who was still so influential in the US nuclear navy, and thus key for continued British access to US nuclear propulsion secrets.

57 TNA AIR 8/2785, David Owen to Prime Minister, Comprehensive Test Ban: Warhead Stockpile Safety and Reliability, 31 March 1978.

58 TNA DEFE 19/241, V.H.B. Macklen, DCA (PN) to PS/Secretary of State, Stockpile Safety and Reliability, 3 April 1978.

59 TNA DEFE 19/241, B.M. Norbury, Head of DS 11 to APS/SoS, Comprehensive Test Ban – Warhead Stockpile Safety and Reliability, 3 April 1978.

60 Lord Zuckerman to Foreign Secretary, Comments on GEN. 63 (78) 4, 30 March 1978 reproduced in David Owen, *Nuclear Papers*, Liverpool, Liverpool University Press, 2009, pp. 227–229.

61 TNA DEFE 19/242, V.H.B. Macklen to PUS, CTBT – Stockpile Maintenance, 8 May 1978.

62 John R. Walker, *British Nuclear Weapons and Test Ban 1954–1973: Britain, the United States, Weapons Policies and Nuclear Testing: Tension and Contradictions*, Farnham, Ashgate, 2010, p. 265.

63 By 1978 the UK nuclear weapons stockpile had reached its maximum level of about 462 warheads – see John R. Walker and John Simpson, *British Nuclear Weapon Stockpiles 1953–1978*, RUSI Journal, October 2011. Of these the Polaris ET317 warhead was by far the largest single element at 191 and had been produced between January 1967 and September 1969. The 111 WE177Cs were the last introduction to the stockpile and these had been produced between 1973 and 1977. This warhead had been produced without any prior underground nuclear explosion as it was based on established designs. See John R. Walker, *A History of the United Kingdom's WE177 Nuclear Weapons Programme*, London, BASIC, 2019, www.basicint.org/wp-content/uploads/2019/08/A-History-of-the-UK-WE-177.pdf; and John R. Walker, 'British Nuclear Weapons Stockpiles by Year: 1953–1977', *RUSI Journal*, Vol. 166, No. 4, pp. 10–20, 2020. As it happens the last WE177A was taken out of service in March 1998 – the last UK underground nuclear test – Bristol – had been in November 1991 with a yield under 20 kilotons.

64 TNA AIR 8/2785, B.M. Norbury, Head of DS 11 to APS/S of S, Comprehensive Test Ban: Warhead Stockpile Safety and Reliability, 3 April 1978.

65 See John R. Walker, *A History of the United Kingdom's WE177 Nuclear Weapons Programme*, London, BASIC, 2019, pp. 23–27.

66 John R. Walker, *British Nuclear Weapons and Test Ban 1954–1973: Britain, the United States, Weapons Policies and Nuclear Testing: Tension and Contradictions*, Farnham, Ashgate, 2010, p. 204.

67 TNA CAB 120/1011, GEN 63 (78) 7th Meeting Cabinet Official Group on International Aspects of Nuclear Defence Policy, 10 March 1978.

68 See John R. Walker, *A History of the United Kingdom's WE177 Nuclear Weapons Programme*, London, BASIC, 2019.

69 TNA CAB 120/1011, GEN 63 (78) 7th Meeting, Cabinet Official Group on International Aspects of Nuclear Defence Policy, 10 March 1978.

70 TNA DEFE 23/218, DCA (PN) 90/78, Safety and Reliability of the UK Nuclear Weapon Stockpile under a Comprehensive Test Ban (CTB), Note by DCA (PN), 13 March 1978.

71 Hydronuclear tests are nuclear weapon tests, or high-explosive-driven criticality experiments, limited to subcritical, or slightly supercritical neutron multiplication. They can be designed to release negligible or at most very small amounts of fission energy. The prefix 'hydro' means in this instance that the core of the nuclear device behaves like a fluid under compression by the chemical high explosive. Sufficient energy may be released to melt the core, but the nuclear energy released is insufficient for the core to heat to plasma temperatures and explode 'like a bomb'. Thomas B. Cochran and Christopher E. Paine, *The Role of Hydronuclear Tests and Other Low-Yield Nuclear Explosions and Their Status Under A Comprehensive Test Ban*, New York, Natural Resources Defense Council Inc., April 1995 (Rev1). The United States historically used a definition of 'hydronuclear' as being less than 0.002 tons (2 kg) of yield. However, another definition for a hydronuclear test involves a nuclear yield no larger than the energy provided by the chemical explosive that

drove the implosion. See National Research Council, *The Comprehensive Nuclear Test Ban Treaty: Technical Issues for the United States*, Washington, DC, The National Academies Press, 2012, p. 102.

72 TNA FCO 66/1080, A. Reeve to C.L.G. Mallaby, ACDD, FCO, 20 March 1978.

73 TNA FCO 66/1080, C.L.G. Mallaby, Arms Control and Disarmament Department to Sir A. Duff and PS/Mr Judd, Mr Jay's Lunch with Dr Harold Brown, 12 April 1978.

74 TNA CAB 120/1011, GEN 63 (78), 8th Meeting Cabinet Official Group on International Aspects of Nuclear Defence Policy, 14 March 1978.

75 TNA CAB 120/1011, GEN 63 (78) 9th Meeting Cabinet Official Group on International Aspects of Nuclear Defence Policy, 17 March 1978.

76 TNA AIR 8/2785, John Hunt to Prime Minister, Comprehensive Test Ban: Warhead Stockpile Safety and Reliability, 20 March 1978.

77 TNA AIR 8/2785, Extract from Cabinet paper on Nuclear Defence Policy Reference A06996, 1. Comprehensive Test Ban: Warhead Stockpile Safety and Reliability, 10 April 1978. See also B.M. Norbury, Head of DS 11 to APS/S of S, Comprehensive Test Ban: Warhead Stockpile Safety and Reliability, 31 March 1978.

78 TNA DEFE 19/214, J.C. Edmonds, Foreign and Commonwealth Office to Sir Clive Rose, Cabinet Office, CTB Negotiations: The Stockpile Problem and the Duration of the Treaty, 19 April 1978.

79 TNA DEFE 19/241, B.G. Cartledge, 10 Downing Street to G.G.H. Walden, Foreign and Commonwealth Office, Comprehensive Test Ban, 25 April 1978.

80 *Foreign Relations of the United States, 1977–1980*, Vol. XXVI, Memorandum of Conversation, Summary of a Telephone Conversation between the President and Prime Minister Callaghan, Washington, DC, 17 April 1978, 2:27–2:47 p.m., pp 464–465.

81 *Foreign Relations of the United States, 1977–1980*, Vol. XXVI, Memorandum from Robert Hunter of the National Security Council Staff to the President's Assistant for National Security Affairs (Brzezinski), Washington, DC, 3 May 1978, Your meeting with British Ambassador Jay, April 21, 12:10 to 12:35.

82 TNA CAB 120/1011, GEN 63 (78), 11th Meeting Cabinet Official Group on International Aspects of Nuclear Defence Policy, 26 April 1978.

83 TNA CAB 120/1011, GEN 63 (78), 11th Meeting Cabinet Official Group on International Aspects of Nuclear Defence Policy, 26 April 1978.

84 TNA DEFE 19/242, Percy Craddock, Head of UK Delegation to the CTB Negotiations, FCO to David Owen, Negotiations for a Comprehensive Ban on Nuclear Tests: Part II (Comment), 21 April 1978.

85 TNA CAB 120/1011, GEN 63 (78), 12th Meeting Cabinet Official Group on International Aspects of Nuclear Defence Policy, 5 June 1978; TNA DEFE 19/242, J.D. Bryars, AUS (Defence Staff) to PS/Secretary of State, Comprehensive Test Ban: The Stockpile Problem, 22 May 1978.

86 TNA DEFE 19/242, W.K. Prendergast, Foreign and Commonwealth Office to B.G. Cartledge, 10 Downing Street, Comprehensive Test Ban, 25 May 1978.

87 *Foreign Relations of the United States, 1977–1980*, Vol. XXVI, Memorandum from the President's Assistant for National Security Affairs (Brzezinski) to President Carter, CTB, Washington, DC, May 10, 1978, pp. 477–479.

88 *Foreign Relations of the United States, 1977–1980*, Vol. XXVI, Presidential Directive/NSC-38, Comprehensive Test Ban, Washington, DC, 20 May 1978, pp. 482–483.

89 Raymond Garthoff, *Détente and Confrontation American-Soviet Relations from Nixon to Reagan*, Washington, DC, The Brookings Institution, 1985, p. 757; see *Foreign Relations of the United States, 1977–1980*, Vol. XXVI, Memorandum From Secretary of Energy Schlesinger to President Carter, Continued Discussion of a Zero-Yield CTB, Washington, DC, 30 May 1978, pp. 486–488; Memorandum From Secretary of Defense Brown to President Carter, Washington, DC, 1 June 1978, pp. 488–493.

90 TNA DEFE 19/242, V.H.B. Macklen, DCA (PN) to Sir Clive Rose, Cabinet Office, CTB – Recent US SCC Meeting, 30 June 1978.

91 *Foreign Relations of the United States, 1977–1980*, Vol. XXVI, Memorandum From Secretary of Defense Brown to President Carter, Washington, 1 June 1978, pp. 488–493.

92 TNA DEFE 19/242, V.H.B. Macklen to PUS, CDS, CSA and DUS (P), CTB – Recent US Developments, 30 June 1978.

93 TNA DEFE 19/242, J.D. Bryars, AUS (Defence Staff) to PS Secretary of State, Comprehensive Test Ban: The Stockpile Problem, 25 May 1978.

94 TNA DEFE 19/242, Clive Rose, Cabinet Office to V.H.B. Macklen, DCA (PN), MOD, CTB: Permitted Experiments, 7 June 1978.

95 TNA DEFE 19/242, G.G.H. Walden, Foreign and Commonwealth Office to Bryan Cartledge, 10 Downing Street, Comprehensive Test Ban, 22 June 1978.

96 TNA DEFE 19/242, V.H.B. Macklen to Sir Clive Rose, CTB-Recent US SCC Meeting, 30 June 1978.

97 The characters of the commedia usually represent fixed social types and stock characters such as foolish old men, devious servants or military officers full of false bravado. Commedia dell'arte – Wikipedia. So perhaps the interagency process here was closer to commedia than Shakespeare. https://en.wikipedia.org/wiki/Commedia_dell%27arte

98 *Foreign Relations of the United States, 1977–1980*, Vol. XXVI, Memorandum of Conversation Summary of Meeting with the President on CTB Issues, Washington, DC, 13 June 1978, 2:05–3:30 p.m., pp. 497–508.

99 *Foreign Relations of the United States, 1977–1980*, Vol. XXVI, Summary of Conclusions of a Special Coordination Committee Meeting, Comprehensive Test Ban (CTB), Washington, DC, 27 June 1978, pp. 509–511.

100 *Foreign Relations of the United States, 1977–1980*, Vol. XXVI, Memorandum From Secretary of Energy Schlesinger to the President's Assistant for National Security Affairs (Brzezinski), Washington, DC, 1 July 1978, pp. 512–513.

101 *Foreign Relations of the United States, 1977–1980*, Vol. XXVI, Minutes of a Special Coordination Committee Meeting Washington, Comprehensive Test Ban (CTB), Washington, DC, 6 July 1978, pp. 514–519.

102 See, for example, Max Hastings, *The Abyss The Cuban Missile Crisis 1962*, London, William Collins, 2022 for an account of the JCS obsession with bombing and invading Cuba.

103 *Foreign Relations of the United States, 1977–1980*, Vol. XXVI, Memorandum From the Director of the Arms Control and Disarmament Agency (Warnke) to President Carter, Washington, DC, 28 July 1978, pp. 532–535.

104 TNA DEFE 19/242, C.D. Verey, DS 11 to AUS (D Staff), CTBT Stockpile Maintenance – Carter Decision, 23 May 1978.

105 TNA FCO 66/1069, M.J. Vile, Cabinet Office, Comprehensive Test Ban Negotiations, 29 August 1978.

106 TNA FCO 66/1473, A. Reeve, Arms Control and Disarmament Department, FCO to D.C. Fakley, ACSA (N), MOD, CTB: Policy Review, Draft Policy Review Paper, paragraph 7, 23 October 1980.

107 TNA FCO 66/1073, D.C. Fakley, Director Defence Science 6 to Sir C. Rose, Cabinet Office, CTBT Negotiations; Permitted Experiments, 22 September 1978.

108 Raymond Garthoff, *Détente and Confrontation American-Soviet Relations from Nixon to Reagan*, Washington, DC, The Brookings Institution, 1985, p. 757.

109 TNA FCO 66/1073, CTB Negotiations: Permitted Experiments, Note by the Ministry of Defence, 22 September 1978.

110 TNA CAB 130/1011, GEN 63 (78), 19th Meeting, Item No.1 Comprehensive Test Ban: Duration and Related Issues, 19 September 1978.

111 TNA CAB 130/1011, GEN 63 (78), 19th Meeting, Item No.1 Comprehensive Test Ban: Duration and Related Issues, 19 September 1978.

112 See Lorna Arnold and Mark Smith, *Britain, Australia and the Bomb: The Nuclear Tests and Their Aftermath*, Basingstoke, Palgrave Macmillan, 2006.

113 See for example TNA DEFE 72/151, Surveillance Annual Report, T.P. O'Callaghan, Post Development Services, Hunting Engineering to A.Arm.13, MOD, 1 August 1973.

114 TNA CAB 130/1011, GEN 63 (78), 20th Meeting, Item No. 1 Comprehensive Test Ban, 26 September 1978.

115 *Foreign Relations of the United States, 1977–1980*, Vol. XXVI, FM The White House Situation Room. To Dr Brzezinski for the President. WH81232. Message to President Carter from Prime Minister Callaghan, Washington, DC, 14 September 1978, 1203Z, pp. 537–538,

116 *Foreign Relations of the United States, 1977–1980*, Vol. XXVI, Message From President Carter to Prime Minister Callaghan, pp. 541–542.

117 TNA FCO 66/1072, C.L.G. Mallaby, Arms Control and Disarmament Department, FCO to P.J. Weston, Washington, CTB: Permitted Experiments, 3 October 1978.

118 TNA FCO 66/1072, M.A. Pakenham, Washington to C.L.G. Mallaby, Arms Control and Disarmament Department, FCO, Permitted Experiments, 10 October 1978.

119 TNA FCO 66/1072, D.C. Fakley to Mr Edmonds, Conversation with General Giller and Mr Duff, 5 October 1978.

120 TNA CAB 130/1011, GEN 63 (78), 24th Meeting, Item No.1 Comprehensive Test Ban: Current State of Negotiations, 6 November 1978.

121 TNA CAB 130/1011, GEN 63 (78), 25th Meeting, Comprehensive Test Ban: Permitted Experiments, 10 November 1978.

122 TNA CAB 130/1011, GEN 63 (78), 18, Comprehensive Test Ban: Permitted Experiments, Note by the Secretary, 23 November 1978.

123 TNA FCO 66/1072, C.L.G. Mallaby, Arms Control and Disarmament Department to Mr Moberly, PS/PUS, Private Secretary, Comprehensive Test Ban, 22 November 1978.

124 TNA FCO 66/1072, G.G.H. Walden to Mr Mallaby, CTB, 28 November 1978.

125 TNA CAB 130/1011, Cabinet Official Committee on International Aspects of Nuclear Defence Policy, Comprehensive Test Ban: Permitted Experiments, Note by the Secretary, Annex, GEN 63, 15, 1 November 1978.

126 TNA CAB 130/1011, Cabinet Official Committee on International Aspects of Nuclear Defence Policy, Comprehensive Test Ban: Permitted Experiments, Note by the Secretary, GEN 63 (78), 18, 23 November 1978.

127 TNA FCO 66/1286, C.D. Verey, MOD to G.G. Wetherell, ACDD, FCO, CTB: UK/ US Bilateral Discussion, 17/18 January 1979 – Brief on 'Permitted Experiments', 12 January 1979. See also TNA FCO 66/1072, John Hunt to Prime Minister, Comprehensive Test Ban: 'Permitted Experiments,' 27 November 1978 and TNA FCO 66/1072, Bryan Cartledge, 10 Downing Street to G.G.H. Walden, Foreign and Commonwealth Office, Comprehensive Test Ban: Permitted Nuclear Experiments, 28 November 1978.

128 TNA FCO 66/1072, Roger Facer, Minister of Defence to B.G. Cartledge, No 10 Downing Street, CTB: Procedural Aspects and 'Permitted Experiments', 1 December 1978.

129 TNA FCO 66/1072, G.G.H. Walden, Foreign and Commonwealth Office to Bryan Cartledge, No 10 Downing Street, Comprehensive Test Ban: Procedural Aspects and 'Permitted Experiments,' 4 December 1978.

130 TNA CAB 130/1011, Cabinet Official Committee on International Aspects of Nuclear Defence Policy, Minutes of a Meeting, GEN 63 (78) 26th Meeting, 5 December 1978.

131 TNA FCO 66/1286, C.L.G. Mallaby, ACDD to B.M. Norbury, DS 11, MOD, CTBT: Permitted Experiments: Brief for UK/US Bilateral, 17/18 January 1979.

132 TNA FCO 66/1286, C.D. Verey, MOD to G.G. Wetherell, ACDD, FCO, CTB: UK/ US Bilateral Discussion, 17/18 January 1979 – Brief on 'Permitted Experiments', 12 January 1979.

133 TNA FCO 66/1286, C.L.G. Mallaby, ACDD to B.M. Norbury, DS 11, MOD, CTBT: Permitted Experiments: Brief for UK/US Bilateral, 17/18 January 1979.

134 TNA FCO 66/1308, Record of a Meeting in the Cabinet Office, Comprehensive Test Ban Negotiations, 16 February 1979.

135 TNA FCO 66/1308, CTB: Discussion with the Leader of the US CTB Delegations, C.L.G. Mallaby, ACDD, FCO to J.C. Edmonds, Geneva, 23 February 1979.

136 TNA FCO 66/1308, CTB: Discussion with the Leader of the US CTB Delegations, C.L.G. Mallaby, ACDD, FCO to J.C. Edmonds, Geneva, 23 February 1979.

137 TNA FCO 66/1269, Comprehensive Test Ban Negotiations, Record of a UK/US Bilateral Meeting Held at the US SALT Building on Friday 2 March 1979 paragraph 29 and Comprehensive Test Ban Negotiations, Record of a UK/US Bilateral Meeting Held in Dr Johnson's Apartment at the Residence de France on 12 March 1979.

138 TNA FCO 66/1269, CTB: Conversation with Dr York, 20 March 1979, J.C. Edmonds, 21 March 1979.

139 TNA FCO 66/1292, Briefing on CTBT, Clive Rose, Cabinet Office to P.H. Moberly, FCO, 25 April 1979. This attached a briefing on CTBT negotiations for an incoming Conservative Prime Minister with copies also going to the Foreign and Defence Secretaries.

140 TNA FCO 66/1269, Comprehensive Test Ban Negotiations, Record of a UK/US Bilateral Meeting at the UK Mission on 17 July 1979.

141 TNA FCO 66/1305, US/UK Consultations on Threshold Test Ban, D. Fakley, ACSA (N), MOD to Mr A. Reeve, ACDD, FCO, 6 December 1979.

142 TNA FCO 66/1310, Comprehensive Test Ban, B.G. Cartledge, 10 Downing Street to M.J. Vile, Cabinet Office, 8 May 1979.

143 TNA PREM 19/212, Note of Part of a Discussion between the Prime Minister and the US Secretary of State Mr Cyrus Vance at 10 Downing Street on 23 May 1979.

144 *Foreign Relations of the United States, 1977–1980*, Vol. XXVI, Telegram From the Secretary's Delegation in London to the Department of State, Section No 4025. Department for Christopher only. White House to Dr Brzezinski for the President. Subject: Meeting with PM Thatcher, London, 24 May 1979, 0129Z, p. 574. This remark was made in the context of salt mines in the USSR that could be used to conceal effectively a clandestine underground nuclear test, but probably also extended to other testing matters too. See UK record at TNA PREM 19/212, Note of Part of a Discussion between the Prime Minister and the US Secretary of State Mr Cyrus Vance at 10 Downing Street on 23 May 1979.

145 TNA PREM 19/212, Note of Part of a Discussion between the Prime Minister and the US Secretary of State Mr Cyrus Vance at 10 Downing Street on 23 May 1979.

146 Dr Frank Panton, former Assistant Chief Scientific Advisor (Nuclear) MOD and a member of the 1958–1962 UK delegation to the Conference on the Discontinuance of Nuclear Weapons Tests. Panton frequently referred to this experience in conversations with the author, noting that the Russians resumed testing at two separate sites, which must have been months in the planning.

147 TNA DEFE 23/1344, Report of a Meeting held in the US State Department on the Afternoon of Monday to Discuss Comprehensive Test Bans, 14 March 1977.

148 TNA FCO 66/1310, Comprehensive Test Ban Negotiations, C.L.G. Mallaby, ACDD to Private Secretary, 11 May 1979.

149 TNA PREM 19/212, Manuscript comments on Comprehensive Test Ban Negotiations, May 1979.

150 TNA PREM 19/213, Note of a Meeting on the Comprehensive Test Ban Treaty Held in 10 Downing Street on Monday, 18 June 1979.

151 TNA FCO 66/1310, Comprehensive Test Ban (CTB), R.L.L. Facer, MOD to B.G. Cartledge 10 Downing Street, 11 May 1979.

152 TNA PREM 19/213, Comprehensive Test Ban Negotiations; National Seismic Stations (NSS) on UK territory, UKMis Geneva telegram No. 346 to FCO, 20 July 1979.

153 TNA FCO 66/1310, CTB: Soviet Probe about a Threshold Test Ban, A. Reeve, ACDD, FCO to J.C. Edmonds, c/o ACDD, 10 December 1979: Soviet Probe about a

Threshold Test Ban, P.J. Weston, Washington to A. Reeve, ACDD, FCO, 17 December 1979.

154 TNA CAB 130/1158, MISC 1 (81) 2, Cabinet Official Group on International Aspects of Nuclear Defence Policy, Nuclear Test Ban Policy: Draft Report, Note by the Secretary, Annex B, Comprehensive Test Ban: Nuclear Advisory Panel Views, 29 January 1981.

155 See for example TNA DEFE 72/151, Surveillance Annual Report, T.P. O'Callaghan, Post Development Services, Hunting Engineering to A. Arm.13, MOD, 1 August 1973.

156 TNA DEFE 13/1768, Nuclear Warhead Research and the capabilities of AWRE, Report by DCA (PN), Annex A to CSA 389/76, 6 September 1976.

157 See David Owen, *Nuclear Papers*, Liverpool, Liverpool University Press, 2009.

158 TNA FCO 66/1473, A. Reeve, Arms Control and Disarmament Department to Mr P.H. Moberly, MISC 1 CTB, 20 November at 3.00 pm, 19 November 1980.

159 The WE177 B was first deployed on September 1966 with its deployment completed by the end of May 1967. See John R. Walker, *A History of the United Kingdom's WE177 Nuclear Weapons Programme*, London, BASIC, 2019.

160 John R. Walker, 'British Nuclear Weapons Stockpiles by Year: 1953–1977', *RUSI Journal*, Vol. 166, No. 4, pp. 10–20, 2020.

161 TNA DEFE 19/240, UK Stockpile Reliability in the Absence of Nuclear Experiments, AWRE, 1 February 1978.

162 *United States Nuclear Tests July 1945 through September 1992*, U.S. Department of Energy, National Nuclear Security Administration, Nevada Field Office, DOE/NV-209-REV, 16 September 2015, www.nnss.gov/docs/docs_LibraryPublications/DOE_NV-209_Rev16.pdf.

163 The Committee on Disarmament, originally established in 1979, became the Conference on Disarmament in 1984.

164 Defence Secretary Des Browne, Hansard, House of Commons, column 1944W, 13 July 2006.

5 Chevaline and Successor Systems – Strategic and Tactical

Introduction and context

As already discussed, the prospect of a Comprehensive Nuclear Test Ban Treaty would constrain, or at worst hobble, British plans to remain in the nuclear business. So plans for further UK nuclear tests were developed alongside UK participation in the 1977–1980 nuclear test ban negotiations. Until such times as a test ban entered into force, the UK would continue to test in order to acquire the maximum amount of design and other information. Chapter 3 looked at UK test plans and objectives from 1974 to 1982 and assessed the extent to which these were fundamental to the UK remaining in the nuclear weapons business for the long term. This chapter looks in detail at the specific weapon systems as they entered into operational service – only two new warheads entered the UK stockpile in the period covered by this book – the WE177C and Chevaline. Finally, the chapter will also offer some insights on the AWRE, MOD and Ministers' initial thinking on a new warhead for the Trident submarine-launched ballistic missile system, in terms of both design and the eventual number of warheads needed for this system. There is limited detail, however, available at The National Archives on these issues, so the picture is somewhat patchy.

Chevaline and a resumption of UK nuclear testing

Once the UK finally decided to embark substantively on the Polaris Improvement Programme (PIP) in 1973, a resumption of underground testing was unavoidable in order to develop a hardened warhead to fit into the new front end of the Polaris missile.[1] The E317 warhead carried on Polaris A3T missiles had an originally assigned life span of 12 years.[2] Seventy per cent of the warheads had been built between 1968 and 1970, which would mean that 1982 would see the end of ET317 in operational service unless its allotted life span could be extended.[3] The new Chevaline warhead for an improved front end would likely have a similar life span as it would re-use some of the same components as incorporated into ET317; for example, Units 19002 were virtually directly incorporated into Chevaline warheads (known as FS646) and the tritium gas bottles were common to both as well.[4] However, by the end of the

DOI: 10.4324/9781003375708-5

1970s AWRE and the MOD also needed to think about a new warhead design to replace the WE177 tactical nuclear weapon family as well as a design for a potential future strategic successor system. Thinking about these options had necessarily to proceed in advance of a firm political decision to acquire new warheads for strategic or tactical forces otherwise there would likely be a much longer lead time in acquiring a new system and the likelihood that the UK's ability to sustain a continued operational capability would fail. Obsolescence and withdrawal would outpace a successor and its deployment. Any new warheads, however, could not be developed and confirmed without two and possibly three underground nuclear tests. A key concern for the UK in this period (the late 1970s and early 1980s) was that although much of the proving of a new warhead could now be done by calculation and laboratory experiment, AWRE still did not know enough about warhead physics to allow a new warhead to enter the stockpile without at least one nuclear test to validate the design.[5] Defence Policy Staff in the MOD had an expectation that once Aldermaston had finished the development of the Chevaline warhead, towards the end of 1977, it would start work towards a replacement programme for the WE177 weapons, which would become life expired in the early 1980s.[6]

The WE177 C tactical nuclear weapon for the RAF

In 1969 the Air Force Department at the MOD was interested in the development of a new warhead design, particularly to the possibility of developing the current production of the WE 177A to produce a medium-yield warhead suitable for use against hardened airfield targets.[7] The operational requirement for this higher-yield warhead stemmed from an original request from SACEUR in 1970.[8] Essentially the WE177A, with a yield of 10 kilotons, was considered too low for certain Warsaw Pact targets in Eastern Europe, such as hardened aircraft shelters, whilst the WE177B at 450 kilotons, was deemed to be far too large for use in Eastern Europe. Instead, the UK was able to offer SACEUR an intermediate-yield weapon, which was designated the WE177C. The WE177C warhead itself was a two-stage weapon using the same ballistic casing as the WE177B, but in order to keep the yield below 200 kilotons, which seems to have been the threshold for tactical use in Eastern Europe, the secondary component (the thermonuclear stage) likely contained a reduced quantity of lithium deuteride 6 fuel for a nominal yield of 190 kilotons. Since the WE177C used the same primary as the WE177A and WE177B, there was no need for AWRE to validate the warhead design with a new UK nuclear test at Nevada. The weapons production campaign itself at ROF Burghfield ran from 1973 to 1977.[9] Overall it had taken the UK some ten years – from September 1966 to May 1977 to complete the production of the three different types of the WE177 tactical nuclear weapon – the A, B and C variants whose genesis goes back to 1959.[10] This would mean that individual weapons' shelf life would expire at different times over a similar period of time, that is, ten years. The WE177C's original design shelf life was 16 years, although its

Operational Requirement had called for 20 years. Both the WE177A and B would also be nearing the end of their shelf lives too, and in the case of the WE177B much sooner as the first weapons had entered service in September 1966. For these reasons, the UK had to start thinking about a replacement warhead and delivery system or systems at roughly the same time as the tripartite test ban negotiations were beginning in Geneva in the summer of 1977. Indeed as we have seen in Chapter 3 (and further discussed in Annex 1), two of the UK's nuclear tests (Nessel, 1979 and Dutchess, 1980) certainly included some elements that would enable AWRE to design a new tactical warhead in future. A Theatre Nuclear Weapons Requirements Committee: WE177 replacement had convened in Whitehall as early as 1977.[11] In 1978 the Theatre Nuclear Weapons Policy Steering Group (TNWPSG) had informed the Chiefs of Staff that the WE177s would need to be replaced from 1986 onwards. Naval Air Staff Target 1231 was subsequently endorsed by the MOD's Operational Requirements Committee (Nuclear) in December 1978. NAST 1231, however, only provided a basis for a nine-month feasibility study of one or more weapons to replace the WE177 and included an examination of the implications of providing yields outside the range of those achieved by the current weapons (0.5 kiloton, 10 kilotons, 190 kilotons and 450 kilotons).[12] A report by the Feasibility Study was completed in March 1980.[13] Given the resource pressures on AWRE, scientists there were not able to create a project definition for a new warhead before the mid-1980s and production before the early 1990s.[14] By late summer 1982 the Theatre Nuclear Weapons Policy Steering Group was still looking into the question of future UK-owned theatre nuclear weapons.[15] Despite all this, substantive discussions and planning on a WE177 replacement would continue well after the period discussed in this book and into the 1990s.[16] In fact, the MOD recognised that given resource constraints at AWRE, it would not be possible to begin the WE177 replacement programme with a new family of weapons before the early 1990s.[17] AWRE would be tied up with the Trident programme until the early or mid-1990s.[18] Planning assumptions in October 1979 noted that the WE177B would not be replaced; instead, it would be refurbished and kept in service during the 1980s and ultimately replaced by a long-range nuclear force. Exactly what this force entailed is not clear from available archival records. In contrast, the W177A and C would be replaced by a new weapon on a one-for-one basis in the early to mid-1990s.[19] In the interim, all three weapons would be refurbished for a five-year life extension.[20] This would mean that the B version would be phased out by 1991, the A between 1994 and 1997 and the C between 1998 and 2002. Development of the new weapon would only begin once the development of the strategic force warhead (i.e. Trident) was completed. There were, in any case, significant staffing problems at AWRE, primarily skilled manpower shortages, which meant that the UK could not conduct more than one new nuclear warhead project at the same time.[21] Decisions on new theatre weapons would therefore have to wait whilst priority at Aldermaston was given to a new strategic system to replace Polaris/Chevaline.[22] In any event, as of the end of

1978, the UK's ability to design and produce modernised and improved tactical nuclear weapons would have been seriously curtailed by the early entry into force of a CTBT, or agreement to any prior participation in a testing moratorium.[23] However, as it transpired, this proved only a theoretical threat as a test ban treaty proved elusive for the reasons explained in Chapters 2 and 4. In any event, the Cabinet's Overseas Policy and Defence (Nuclear) Committee decided in June 1993 that the UK would not proceed with a successor to the WE177 and that it would plan to rely on Trident in a sub-strategic role as a replacement capability.[24] Finally, old age and the end of the Cold War put paid to the UK's theatre nuclear weapons as the last four WE177As were withdrawn from operational service on 22 April 1998 from RAF Marham.[25]

ET317 and Chevaline warheads (SF 646)[26]

The UK conducted three nuclear tests in 1974, 1976 and 1977, and as a result of these AWRE now had a proven hardened warhead for Chevaline.[27] Since about 1967 AWRE had directed much of its R&D capability to the development of such a warhead. By July 1977 it was a proven design and the production was tooled up and ready to begin manufacture.[28] The overall cost of providing hardened warheads within the Chevaline programme was £105 million, of which £51 million had already been spent with the remainder mostly going on production as development work was practically complete.[29] Chevaline's hardened warhead used slightly more plutonium than the ET317 in the original Polaris system.[30] A sophisticated combination of hardened warheads and decoys was the principal feature of Chevaline.[31] Apparently, this new warhead design posed severe safety problems since the explosive (EDC 32) used in the implosion system was more sensitive than that used in the ET317 warhead. On the other hand, the new warhead had superior nuclear safety as it was inherently one point safe, thereby enabling the mechanical safing arrangements in ET317 to be dispensed with. AWRE's calculations, validated by US tests, were used to assure the one-point safety of the Chevaline warhead.[32] It is worth stressing here that this shows that the UK weapons programme also depended critically on data derived from US nuclear tests provided under the 1958 Mutual Defence Agreement, which certainly helped reduce the need for additional, and expensive, full-scale British tests.[33] In 1975 the key planned target dates for Chevaline production and deployment were as follows:

Acceptance of First Demonstration and Shakedown Operation Outload July 1979;
1st submarine Tactical outload December 1979;
2nd Demonstration and Shakedown Operation Outload April 1980;
2nd submarine Tactical outload September 1980;
3rd Demonstration and Shakedown Operation Outload;
3rd Submarine Tactical outload March 1981;
Completion of Tactical spares/surveillance rounds March 1982.[34]

Despite this schedule, the Ordnance Board had still to approve the design as of April 1976.[35] The relationship between planned dates and their implementation in the Chevaline programme was always a loose one, which is a reminder that nuclear weapons programmes do not always run smoothly as unforeseen contingencies are apt to derail or put extra pressure on timetables. And this is before the impact of any potential arms control measure, such as a CTBT, is taken into consideration.

Battler

Efforts were also underway to develop a lighter warhead for Chevaline as noted in Chapter 3. This was codenamed Battler and would have helped extend the operating range of Polaris thereby giving the Resolution class submarines more sea room in which to hide from the depredations of Soviet anti-submarine efforts. It does not seem to be clear from available records whether this desire for extended range emanated from a Royal Navy operational requirement, or whether it was simply a prospect opened up by advances in AWRE's warhead design capabilities. Available evidence suggests that this was more the result of the AWRE's efforts rather than Admiralty requirements (see also Chapter 3). That said, concerns over the threat to the UK deterrent that future major advances in Soviet ASW capabilities could pose might have been a factor.[36] To set the context we need to recap on the testing programme connected with Chevaline. As noted in Chapter 3, there was a balance to be struck between the 64 nautical mile increase in range, the reduction in warhead yield that this entailed and the prospective costs of the modification.[37] Questions were raised inside the MOD in late summer 1977 as the tripartite test ban negotiations began whether a testing moratorium in 1978 might adversely impact on the Chevaline programme and the underground test planned for the spring of 1978. However, the UK's assessment seems to have been that the test ban treaty negotiations were moving too slowly to make real a risk that any moratorium might prevent the test originally planned for March 1978.[38] Although these plans for a lighter warhead were in train, a moratorium or test ban would not per se damage the Chevaline programme, or require its cancellation albeit at the expense of precluding making a relatively minor modification that might be judged worthwhile because of its effect on Polaris' range.[39] There could be an increase beyond the 1,950 nautical miles set for the proven hardened design.[40] As part of this project, a lighter warhead codenamed Findhorn was tested in April 1978. MOD's original plan had envisaged the introduction of Battler-configured payloads for Chevaline from October 1983 for the third Demonstration and Shake Down Operation (DASO) and all subsequent DASO and tactical outloads.[41] A key point to note here is that the Admiralty Board had agreed back in February 1975 that from then onwards it would be sufficient to maintain three complete tactical outloads plus one spare partial tactical outload of missiles for Polaris A3T and A3TK Polaris missiles. This was a key decision with significant implications for the size of the UK strategic stockpile.[42] Naval staff had concluded that it would be justified in abandoning the capability of loading all four SSBNs in an emergency rather than incurring substantial expenditure for that purpose; moreover, there was also no need to maintain missile stocks to safeguard the force against a crash modification programme. Such a decision would lead to savings in storage, maintenance and testing facilities, and would also enable fissile material to be diverted for use in the Chevaline programme, thus making it unnecessary to acquire new fissile material for this purpose. Savings here would be substantial.[43] Moreover, this was going to be the only way in which fissile material could be provided for the Chevaline programme as it would be virtually impossible to obtain an equivalent quantity from elsewhere in the programme's planned timescale.[44] This became the basis on which all Chevaline planning had proceeded and would apply to the withdrawal programme for A3T warheads and for the Chevaline warhead production programme at ROF Burghfield.[45] MOD had all along planned to use available reserve stocks of fissile materials, together with materials recovered from ET317 warheads, for the Chevaline warhead programme.[46] All this was managed in such a way as to enable the Chevaline production programme to proceed without the need for any new fissile material. The cycle of old warhead withdrawal and refabrication was a long one, and although with the target programme as agreed, the total stocks of fissile material and components were thought to be sufficient to maintain the production flow of Chevaline warheads, it was still a complex problem of coordinated timings that would clearly require careful management to ensure that all the right decisions were taken at the right time. Production was due to begin in mid-summer 1976 'in the warhead area where some of the designs were sufficiently far advanced for this to be done'.[47] Work on hardening the warhead was virtually complete by the end of September 1977.[48] And this was before the second Chevaline-related test planned for that August, so some components at least could be produced before the testing programme had been completed. The peak load of manufacture of Chevaline's thermonuclear lithium components at the Royal Ordnance Factory Chorley was planned for completion by the end of 1979.[49] Overall the plan at the end of November 1980 had been to complete the Chevaline warhead production run in mid-1985.[50]

However, by late 1977 and into early 1978, resource problems caused Chevaline's programme management to conclude that it was now impracticable to provide fully approved Battler equipment at the required date and at the same time to hold the Chevaline programme to its timetable. For this reason, Chevaline's Chief Weapon System Engineer reluctantly concluded in early January 1979 that it would be impracticable to proceed with the Battler modification without unacceptable risk to the completion of the overall Chevaline programme.[51] The addition of an extra 56 nautical miles (lower than earlier figures) on the range at a cost of some £30 million extra was not worth the operational advantage; moreover, as it could endanger the main Chevaline project, the modification was simply not worth it, so the Defence Secretary was happy to concur with the cancellation, especially as overall programme costs were escalating.[52] There seemed that there might have been a remote possibility that consideration would be given at a future date to introducing the Battler modification during the first warhead refurbishment period, but it is not known for certain from the open files at The National Archives whether this ever occurred. Given the pressures on the future Trident programme, as discussed later, such an eventuality seems unlikely. Despite this cancellation, the MOD still believed that the nuclear test at which the lighter warhead was proved was entirely justified and fruitful on wider grounds.[53] Although these 'wider grounds' were not specified, we can reasonably assume that it would have been the data generated that would have added to AWRE's stock of knowledge and experimental data. Such attributes would be important if a CTBT were to enter into force, and possibly too for a new tactical warhead.

Chevaline: production delays and a refurbishment of ET317?

Even without the added complication of Battler, the Chevaline programme quickly ran into trouble caused by a combination of technical challenges with the penetration aid carrier, the decoy system and production problems at AWRE caused by the temporary closure of the plutonium pit production line in Building A1.1, serious staff shortages and industrial action.[54] AWRE faced a shortage of craftsmen, which meant that plutonium production had to be planned on a campaign rather than continuous basis – a change that inevitably reduced the throughput of fissile components at AWRE for Chevaline warheads from 36 to 18 per year. At the then-present level of craftsmen available at Aldermaston, it was impossible to re-open the plutonium scrap recovery plants, which meant that additional supplies of plutonium – a further 200 kgs of plutonium – would be needed to maintain even the lower rate of production. Unless the wastage of AWRE craftsmen was stopped and then reversed, Aldermaston would find it impossible to meet the then-Chevaline warhead production plans.[55] Indeed the problem only became worse as by early July 1979, AWRE's deputy director noted that the run-down of craftsmen had continued since March, so even the lower level of anticipated warhead component production was in jeopardy.[56] Clearly, this industrial staffing situation was much more of an immediate existential threat to the UK's nuclear weapons programme than a test ban treaty. And to compound matters, all Chevaline production works at AWRE had ceased by early July 1979 as a result of industrial action by the Institution of Professional Civil Servants (IPCS) that affected staff dealing with safety (primarily health physics) and plant and equipment maintenance amongst other things.[57] Given the combination of key staff shortages and industrial action at both AWRE and ROF Burghfield, this meant that the Chevaline warhead programme was now going to be some six months late for 40 warheads (end of June 1981 being the original target) and 12 months late for 78 warheads (end of October 1982).[58]

By September 1979 the Director of Atomic Weapons Production and Facilities (DAWP & F) in the MOD estimated that if plutonium pits started to become available before the end of 1979 to a total quantity of between 24 and 28 by the end of 1980, tactical re-entry bodies could be produced to that number by the end of March 1981. This meant that it would be possible to produce the Acceptance/DASO outload of ten sets of equipment by the end of November 1980.[59] As for subsequent tactical outloads, given that plutonium pits beyond the 24–28 planned for above would not be available before January 1981, the planning assumption was that tactical REBs with all other equipment to match would be available three months after plutonium pit deliveries to ROF Burghfield, to a maximum of three per month. This applied to the second, third and fourth boatloads of missiles. Originally planned outloads for the submarines could thus not be met given the plutonium and staffing problems at Aldermaston. Production delays caused by the shortage of staff at Aldermaston and Burghfield as well as a shortage of fissile material made it impossible to hold to the original outload dates. Moreover, there had also been continuous slippage in the forecasted production rate of re-entry bodies (REBs) needed to house the nuclear warheads.[60] By November 1979 the new planned date for HMS Renown to be outloaded with Chevaline had now moved to June 1981.[61] However, this would only be a partial outload and was contingent on reversing the loss of staff at Aldermaston and Burghfield, the US agreeing to provide fissile material, and that the plutonium production building (A1.1) at AWRE coming back on stream after being shut down for safety reasons for three and a half years following the contamination of some of its workforce. HMS Renown would thus only receive her full outload in November 1981. HMS Revenge would deploy with a full load of Chevaline warheads in early to mid-1983. To say that entry into service dates was a moving target would be an understatement. Maintaining even a minimum deterrent at this time was clearly an effort, notwithstanding US assistance under the MDA.

Despite all these revised plans and forecasts, production of Chevaline re-entry bodies and decoys at AWRE and Burghfield remained a concern to Ministers and officials well into the spring of 1980. Since 1 November 1979, there had been some improvement in the staffing situation, but the UK was still not out of the woods. Several conditions would have to be met to ensure that the programme for the initial outload and deployment for Chevaline could be met.[62] These were: that this improvement in staffing levels was sustained; that there were no further technical or industrial problems to hold up production; that the arrangements now in hand to enable the loan of fissile material from the US were confirmed; and that the plutonium manufacturing line at A1.1 at Aldermaston was brought back on stream by the end of 1980. In the event, the A1.1 production building only came back on stream in January 1981.[63] In fact, AWRE's problem with its plutonium production facilities evidently posed far more of a threat to the Chevaline programme and any future replacement system than a test ban treaty, which by even late 1978 had been looking less and less likely as noted in Chapters 2 and 4. By the end of 1980, the staffing problems at AWRE were now much improved. The approved establishment of health physicists at or above the grade of higher scientific officer or equivalent remained at 38. The number in post by December 1980 has risen to 24, compared with 19 in June of that year and the figure of 7 given in the 1978 Pochin report into radiological health and safety at AWRE.[64]

Given the threat posed by this staffing and facilities problem, MOD officials became obliged to look at whether the Polaris ET317 warhead could be kept in service beyond its initial 12-year shelf life in the event that Chevaline ran into insuperable difficulties and could not be deployed in the planned timescale. The date of the first scheduled patrol had now slipped into 1982. AWRE's original intention of fully refurbishing the ET317 warhead had been abandoned when its succession by Chevaline had become a definitive plan.[65] The first warheads manufactured in 1968 would have been due for their first refurbishment in 1980. If it were to be decided to refurbish ET317 after all, assumably because of significant delays in the Chevaline programme, it would be essential to seek AWRE's view on the feasibility of a life extension or even to decide on a refurbishment/life extension programme. The judgement required in this sort of scenario was precisely why AWRE and MOD experts believed that periodic nuclear testing was required; namely to underpin the scientific and technical judgements of the scientists and engineers on issues such as stockpile safety and reliability of which warhead refurbishment and replacement of components would be a good example. Any decision on ET317 would pose complex challenges in several different areas; for example, some of Burghfield's highly specialised manufacturing processes had not been used since 1974, the Arming and Fuzing Device (AFD) and other warhead parts had been purchased from the US and were no longer being produced there; and, moreover, many of the components were based on out-of-date technology. Furthermore, plans to reuse a limited number of warhead components and to recycle and reuse the plutonium and HEU (Highly Enriched Uranium) were based on a withdrawal of ET317 warheads from service with new Chevaline warheads on an outload-for-outload basis. For this reason, there were thus distinct advantages to the earliest possible return of ET317 warheads to Burghfield for dismantlement.[66] In order to keep the Polaris A3T weapons system a viable option in the event of insuperable difficulties with the completion of the Chevaline programme, the Chief of the Polaris Executive (CPE) felt that it was essential to have a credible, comprehensive plan showing the financial, programme and resource utilisation aspects of a decision to refurbish ET317 beyond its then-present life span.[67] In particular, CPE wanted to know whether a significant extension of the life of ET317 warheads was possible and whether a fall back plan showing the key events involved in refurbishing ET317 warheads could be prepared by the end of February 1980. Unfortunately for CPE, a brief response was not possible given all the other studies required at AWRE, so he had to make do with an interim response.[68]

As it transpired, the then currently approved life for the complete ET 317 warhead was 12 years, not 16 has had originally been supposed by the Director of Atomic Weapons Production and Factories (DAWP&F). This life span was dictated by the estimated lives of the external neutron initiator, which was replaceable, the Fuzing and Firing Unit of which the Arming Firing Device

(AFD) was of US origin and as such unlikely to be replaceable and the condition of the plutonium pit. Work was put in hand at AWRE to examine the possibility of extending the pit and AFD lives to 16 years. If there were to be any life extension beyond 16 years, such a decision would need to be based on three key considerations. First, confirmation would be needed that no results from shelf life testing, surveillance or service withdrawal during recent years invalidated the life expectancy of 16 years.[69] Second, extrapolation of the experience from shelf life testing to determine whether an extension of life of the overall assembly beyond 16 years would be possible, and the formulation of additional testing to give adequate confidence where necessary. We cannot be clear from the open archive at Kew whether this meant testing of individual components or a full nuclear test. We can, however, be reasonably confident that no nuclear test for this purpose was conducted between 1980 and 1982; see Annex 1. Third, confirmation that items with lives of less than the extended life could be maintained during that time. Crucially, it would be deterioration within the nuclear components themselves that could lead to a requirement for a deep refurbishment. AWRE, therefore, examined the possibility of an ET317 life extension in this context. Aldermaston scientists thought that there might be some prospect that it might in the end prove possible to agree on an extension for ET317 beyond 16 years, possibly out to 20 years, subject to the necessary withdrawals and examination proving satisfactory – a requirement which could amount to a further four warheads being taken from the stockpile for this purpose. In short, an extension to 16 years was considered possible and was even expected to be approved formally sometime between March and May 1980. Extension beyond 16 years seemed possible, but that would be dependent on there being no need for deep refurbishment, that is, the replacement of any of the more specialised items for which provision had not already been made. In any event, as was the case with any warhead life expectancy studies, a provisional extension would depend on continuing satisfactory surveillance programme results. AWRE could not, in February 1980, give any guarantees for life beyond 16 years when no ET317 warhead was over 12 years old.[70] For CPE, since AWRE could not provide a firm forecast of the date on which ET317 warheads would become due for refurbishment, there could therefore be no confidence that a critical situation would not arise at a time when the Royal Navy was still having to deploy these warheads at sea.[71] Such uncertainty was incompatible with the requirement to provide an assured second strike capability. In light of these assessments, the challenge for the MOD was now to see how to combine a need to meet the Chevaline warhead production schedule whilst simultaneously keeping the ET317 at sea on top of the Polaris missile.

After studying the matter further, AWRE concluded that deploying A3T again in HMS Renown on its next deployment would delay the first return of ET317 warheads for the recovery of plutonium and Units 19002 (possibly the thermonuclear assembly in the secondary) for reuse in Chevaline by approximately one year.[72] This would not affect the production of the minimum number of Chevaline warheads initially necessary to sustain three outloads, but completion of the remainder of the planned stockpile would be affected in three ways. First, a slight interruption in warhead production would be inevitable late in 1983 as a result of a temporary shortage of plutonium. Second, if A3T were kept in HMS Renown for more than a half-deployment, it would not be possible to repay the US plutonium loan on time, or there would be a lengthy hiatus (roughly two years) in production of Chevaline nuclear assemblies. Third, some 10–20 Unit 19002s would have to be built to sustain production. In the event of any A3T re-outload extending beyond the offload of HMS Repulse in early 1984, careful logistic planning to ensure that the oldest ET317 warheads were withdrawn would be essential to prevent an unacceptable proportion being life-expired and therefore having uncertain reliability. In this event, resumption of ET317 neutron generator production might become necessary.

If no decision on redeployment were forthcoming before the end of 1980, AWRE reckoned that it would prove impossible to have an adequate stock of 'new' ET317 tritium gas reservoirs to support an A3T outload in HMS Renown in June 1981, and exceedingly difficult even if stockpile units with only a fraction of their shelf life remaining were used. Tritium contingency stocks would be seriously eroded, and it could prove necessary to empty Chevaline gas bottles already filled during the latter half of 1980. Although the empty reservoir was common to both ET317 and Chevaline warheads, it would, nevertheless, be necessary to ensure adequate (empty) stocks. Given all of this, it appeared to AWRE that there would be extreme difficulties in considering any proposal for a further deployment of A3T into HMS Renown on the then-present planning date of 30 June 1981 if the decision were not taken until the end of 1980. However, such a proposal would probably be practicable if a decision was made by the end of September 1981. Any proposal to place A3T into HMS Revenge again for its next patrol after refit, however, would be seriously constrained by the ageing ET317 stockpile, and could have a serious, but not accurately quantified at that point in time, effect on Chevaline plutonium pit production.

The immediate impact of any decision to deploy A3T missiles on HMS Renown for its next patrol (or, more tentatively, on HMS Revenge, whether Renown was carrying A3T or A3TK) would be to delay the planned release – assumed to be one-third of the stockpile of 173 ET317 warheads in August 1981.[73] Then current plans called for the breakdown of warheads to release both the nuclear assemblies for recovery of the plutonium for the manufacture of Chevaline nuclear assemblies, and of Units 19002 for virtually direct incorporation into Chevaline warheads. AWRE had sufficient stocks of plutonium (including an expected loan from the US) to manufacture a total of approximately 105 Chevaline warheads, a figure which included the 24 already produced, but the last 30 of these would use material corresponding to the repayment of plutonium that needed to be made to the US. That repayment, however, was dependent on material recovered from the re-working of the high proportion of process scrap and/returned ET317 assemblies. Production of the remaining planned 30 or so assemblies required for the Chevaline stockpile was dependent on the further reworking of scrap or returns. Even with all the expected recovery plants working, if no assembly returns did not start until after HMS Resolution returned its missiles and warheads in mid-1982, there would be a minor interruption in late 1983 before the last batch of plutonium was used to complete the stockpile. All this suggests that the ET317 and Chevaline primaries contained no HEU since one might reasonably expect some reference to supplies of HEU if it were. Furthermore, it seemed that the return of a second off load of missiles would have had to follow immediately on completion of returns from the first off-load if sufficient stocks were to accumulate for repayment of the US loan by the agreed date. Otherwise, the plutonium programme would be interrupted for about two years while the repayment batch was being collected, although fewer than ten warheads would be outstanding.

The Royal Ordnance Factory at Burghfield could break down ET317 warheads down at a rate of up to four and half per month which, allowing for one-third of Units 19002 being declared unsuitable for re-incorporation, could sustain the production of three Chevaline warheads per month. Moreover, there was sufficient nuclear material available throughout 1982, which provided a capability to manufacture up to 20 19002 units if required in addition to the existing stock (of 30 recoveries and 29 new make) to sustain production until further ET317 returns commenced.[74]

As already noted, the preponderance of the ET317 stockpile had been first assembled between 1968 and 1970. If a life extension to 16 years was fully approved, then a programme of re-outloading these warheads that ran into 1984–1986 would clearly be entering a critical time bracket. AWRE had noted that any extension beyond 16 years could only be supported by critical examination as certain warheads neared that age, so long-term predictions were not possible in 1980. We cannot determine for sure whether in fact any of this contingency planning for re-outloading of ET317 warheads was necessary in practice. We do know that the final four Chevaline tests (flight trials), including the Service Acceptance Round, in February 1982 were wholly successful;[75] and that the first Polaris patrol armed with Chevaline began in late 1982 with HMS Renown, which was followed by HMS Revenge, so that by the end of summer 1983 the UK would have for the first time the capability to keep a Chevaline equipped submarine at sea at all times.[76] HMS Resolution and HMS Repulse came next in 1985 and 1987 after refits.[77] All this scheduling implies that life-extended ET317 warheads were not required to remain in operational service. In this context, we should recall that the Admiralty approval had been given for the provision of A3T missiles for three full boat loads (i.e. 16 missiles) plus a partial fourth outfit such as could be obtained from RNAD Coulport's handling/processing margin. In effect, this reduced the minimum requirement of missiles by one boat outfit from 74 to 58 and consequently allowed the stock of warheads (three per Polaris missile) to be reduced to 174. Subsequent approval was given for a similar provision for the A3TK Chevaline missiles and the retention of this stockpile through to 1994.[78]

This extremely complex and convoluted saga of contingency planning and uncertainty highlights the constraints that a nuclear weapons programme faces in a medium power such as the UK where resources and expertise are limited. Moreover, add in the engineering and coordinated timetabling complications, then it seems clear that maintaining an effective operational nuclear deterrent force presents considerable logistical, planning and programmatic challenges, which are not easily mitigated. All this had to be dealt with regardless of what was happening, or rather not happening after 1980, in Geneva at the tripartite test ban talks.

Trident C4 and D5 and the Holbrook warhead

MOD's basic planning assumption at the end of 1978 held that if a CTBT were in force, further major UK warhead development would be ruled out during its duration. If a Polaris successor system were required, then there would only be a warhead design available based on knowledge acquired before mid-1980. (This must have been an allusion to the Quargel, Nessel and Colwick tests; see Chapter 3 and the Annex.) However, the UK might be able to have access to US warhead design data, especially if there were to be cooperation on a new delivery system. Such data would be valuable for validating any new UK design, but would still rule out the UK being able to adopt a newer more radical design.[79] For a successor to Chevaline that would be capable of defeating improved Soviet ABM defences around Moscow, it was going to be essential for AWRE to develop a warhead that would be compatible with a high-speed re-entry body rather than a slower blunt-ended re-entry body such as that used on Chevaline.[80] Information obtained in the Chevaline programme indicated how it would be possible to reduce the size and weight of the warhead without any appreciable loss of yield in order to achieve this objective. Two nuclear tests were conducted in 1978 and 1979 to test the validity of these ideas; these tests, as noted in Chapter 3 and Annex 1, were successful and put the UK well on the road to designing a new warhead for Trident *before* the formal decision by the Conservative government to acquire the C4 SLBM had been taken in 1980. By May 1979 it seems that the UK had three choices: copying a US design, use a design based on a device scheduled for testing in mid-1980, or opt for a design already tested.[81] One of the key considerations in the choice of a future strategic warhead was that it was unlikely that AWRE could produce exact copies of any US design that might be provided on an adequate timescale because of the lack of appropriate production methods in the UK. For this reason, the most probable option would be to adopt a UK design.[82] Britain faced problems in duplicating the advanced warhead fabrication techniques used by the US at that time for their warheads and confidence in a UK production route could be low if it were not proven first by a nuclear test. For a UK design, AWRE would make use of the data derived from the UK tests held to date and from the one planned for mid-1980. Aldermaston's aim was to gather information needed for very small and hardened warhead designs – the 1980 test would provide data for the sort of yield desired as the two previous tests were of designs that would produce a yield lower than that preferred for a strategic successor system. However, as of January 1980, the UK would need to make a final choice between the two warhead designs then under consideration. In making this choice much would depend on balancing a range of factors, some of which could not be assessed until the UK had conducted detailed technical talks with the US on the Trident missile and had acquired some warhead components for experimental purposes.[83]

Such a new warhead would require a major programme involving a project definition study, a detailed engineering phase, a four-year development programme and some three years of production during which the production rate would be built up before the first outload of re-entry vehicles would be ready for the first deployment. The total programme would amount to ten years.[84] A test ban treaty that entered into force before the completion of the then-planned UK nuclear tests programme in August 1980 would have seriously affected confidence in the efficiency of UK warheads to be deployed on the C-4 missile.[85] It seems that the UK assumed that if a test ban treaty were to be agreed, then the available warhead designs based on knowledge acquired before mid-1980 would be the only ones accessible.[86] A test ban treaty that entered into force in 1981 and lasted three years, however, would not prejudice adoption of a new design for Trident as nuclear testing could resume in 1985. That said, the MOD believed that the window for doing this would still be small if a new system were to be ready for deployment in the early 1990s at the end of the planned life span of Chevaline/Polaris.[87] Britain's public line on a CTBT and its compatibility with the Trident weapons programme, however, was made clear in an MOD Open Government Document explaining the rationale behind the decision. Conservative government Ministers felt the need to make the case publicly given the perceived impact and influence of the domestic anti-nuclear movement.[88] Defence Open Government Document 80/23 stated that the UK continued to support the conclusion of a Comprehensive Test Ban Treaty; and, moreover, nothing in the requirements for the new Trident force would lead the UK to modify its support for a successful outcome to the tripartite negotiations as soon as practicable.[89] In view of the state of the negotiations in the summer of 1980, as we have seen in Chapter 2, this statement might appear to an uncharitable mind as being just a shade disingenuous. 'As soon as practicable' was clearly only going to be a long-term prospect. As it turned out AWRE fired three validation tests of the Trident warhead, with the last being in 1986.[90] This latter test – the Holbrook III device – following the series of capability shots over the preceding five years completed the modernisation programmes for the deterrent.[91] This final shot would have been the Darwin test conducted on 25 June 1986.[92] These validation tests served to provide the essential basis on which to demonstrate the suitability of a specific warhead design for service deployment and for validating warhead performance. AWRE also had good access to US designs during this period that no doubt further facilitated UK calculations and assessments.[93]

An expectation in the MOD in 1980 was that production of UK warheads for Trident would be unlikely to begin before the full Chevaline warhead production campaign was finished. On the then-current programme assessment, Chevaline warhead production would be complete in mid-1985. For this reason, the first tactical outload for Trident would be unlikely to be available until 1990.[94] As things transpired, the first Trident warhead was assembled at Burghfield in September 1992 with the first Trident patrol starting in December 1994 – 12 and 14 years after the Conservative government's July 1980 decision. The warhead used for Trident D5 was the same as the one originally designed for the C4.[95] Trident D5 only had a yield half of that of the original Polaris ET317 – 200 kilotons, which gives a yield of 100 kilotons for Holbrook (the code name given to the warhead).[96] It also seems that Trident had a lower yield than the Chevaline warhead.[97] MOD's original intention had been 'to deploy no more than 480 warheads'.[98] This number referred to the total outload of 12 missiles per SSBN with 10 warheads per missile,[99] which would not include an approximately 10% spares for maintenance purposes, which had been the practice with the Polaris, Chevaline and WE177 weapons. That would make for a total stockpile of about 528 warheads, but it is not clear from archival sources whether that was the actual figure built. However, it does seem clear that by mid-November 1981 cost figures for the Trident D5 programme assumed that in the first instance the UK would fill 12 tubes on its SSBNs with missiles each carrying ten warheads, which with a 10% margin also results in 528 warheads.[100] One archival source noted that the programme for C4 MIRV warheads would total 576 beginning in 1988.[101] A set of measures, however, had been proposed to enable the production of warheads at a rate of 100 a year to begin in the middle 1980s. A production rate at this level was deemed essential by the MOD for a successor system alone, without taking into account other warhead production such as for a replacement for the WE177. In contrast at this time, there were only 140 Polaris warheads, so this planned expansion represented a considerable increase in the UK's strategic nuclear arsenal. However, MOD planners took the view that the precise numbers of missiles and warheads that would be required to provide adequate margins for maintenance, tests and trial firings still needed further study. As of mid-January 1983, no decision on the final number of missiles and warheads had been taken by Ministers, but there was an effective ceiling of approximately 500 warheads resulting from the commitment that the move from Trident C4 to D5 would not change significantly the number of warheads deployed by the UK.[102] It seems that the proposed D5 force as of January 1982 was costed on the basis of 12 launch tubes filled, each missile carrying 10 warheads for 120 warheads per SSBN.[103] Essentially the same number of warheads could be deployed if only nine tubes were filled, each missile having a maximum of 14 warheads, that is, 126 warheads. The saving on missile numbers to be acquired would be 11 after allowing for the smaller number of missiles needed to outload the SSBNs and the reduced depot handling margin. However, an extra 74 warheads would be needed to meet this requirement. An absence of an agreed assessment on deterrent criteria appears to have been a key factor here as the Joint Intelligence Committee (JIC) was still unable, as of autumn 1983, to provide an assessment of Soviet perceptions of 'unacceptable damage'.[104] By October 1983, the MOD's then-current thinking was that the missile warhead mix for Trident was likely to lead to outloads of significantly below 128 warheads per boat.[105] However, we can be sure that whatever mix was chosen, the Trident programme placed major demands on fissile material.[106]

Habitual shortages of fissile material had been a periodic feature of the British weapons programme since the 1950s.[107] Consequently, the UK had been heavily dependent on the US for its supplies of such materials since then and would be unable to avoid such dependence for future programmes unless it were willing to incur substantial costs to re-establish indigenous production at British Nuclear Fuels Limited's (BNFL's) Calder Hall and Chapelcross Magnox reactors or elsewhere.[108] An initial MOD enquiry to US DOD and DOE officials in early June 1981 seeking increased US/UK collaboration on the procurement of fissile material for the defence nuclear programme proved 'conspicuously successful'.[109] US officials pointed out that if the President were to endorse such enhanced cooperation, that would improve the planning process and greatly facilitate progress. Accordingly, in September 1981 the Prime Minister wrote to President Reagan in order to ask for US help with supplies, picking up an earlier conditional offer (subject to US needs being met first and the UK doing all it could to meet its needs itself) from President Carter. The point had now been reached where it was certain that the UK would need to procure significant quantities of HEU, weapons-grade plutonium and tritium from the US Department of Energy at various rates from 1985 for some ten years in order to implement UK's deterrent plans.[110] From available records at The National Archives, it would appear that UK requirements were in the order of about one ton of plutonium and ten tons of HEU for weapons purposes.[111] These needs were outlined in July 1979 – a year before the decision to acquire Trident was publicly announced. By late spring 1982, however, BNFL had found a way to reduce the costs of fissile material production, and the Defence Secretary recommended that one and a half tons of the plutonium needed for the Trident programme should now be sourced from BNFL with the remainder coming from the US.[112] BNFL had eight reactors available to support the defence nuclear programme, though two of these were used for tritium production. All of the remaining six could only just meet the UK plutonium for weapons requirement, with no margins for any breakdowns, hence the need to rely on the US for some of the UK's requirement.[113] So the UK had moved from a position of total dependence on the US for its fissile material, to one where the majority of its requirements by value would now be produced in the UK. President Reagan had responded positively in January 1982 to the Prime Minister's September 1981 request for assistance with fissile material supplies for the UK's nuclear weapons programme. However, this had been agreement in principle and was contingent on the actual state of US supplies in relation to the timing of UK requirements.[114] At that stage, in early 1982, the UK was still not in a position to refine the specifications of quantities of strategic nuclear material (SNM) it needed, and the US still had to take its own decisions about its own strategic systems and SNM production programme.[115] In terms of volume, it seems that Britain appears to have been dependent on the US for nearly half of its plutonium as of June 1982.[116] Availability of fissile material, or rather its unavailability in the right quantities and types at the right time, thus also posed a greater risk to the weapons programme than the test ban treaty, which in any case by summer July 1982 had been put off as a longer-term goal for arms control policy. It was just as well that the UK was not also facing international negotiations on a treaty on the cut-off in the production of fissile material for nuclear weapons purposes.

As we saw in Chapter 3 and noted earlier, AWRE and MOD contemplated two different designs for the Trident warhead before finalising on Holbrook. It is not entirely clear from the available papers in TNA when this was finalised and the decision taken. But it does seem probable that Holbrook was the more advanced of the two designs under development and consideration in the late 1970s.[117] So even from this minimum amount of information, it is apparent that the threat of a test ban treaty did not impact adversely on the development of the Trident warhead. There was therefore nothing comparable in the 1977–1980 period to the testing moratorium pressures that closed in on British efforts to finalise its first thermonuclear weapon design in 1957 and 1958 and which required an accelerated development and testing programme thereby adding considerable strains and stresses on AWRE scientists and engineers. Nevertheless, the maintenance and planned development of the UK nuclear weapon programme in the 1980s was still dependent on there being no cessation of nuclear testing except for a three to five year period from 1982 to 1986. However, a ban on testing that lasted longer than five years would not only frustrate the development of the Trident warhead, it would progressively undermine confidence in the viability of existing weapons in the stockpile.[118] It therefore bears repeating that testing was not just about testing and validating new warhead designs, it was also about maintaining expertise.

Conclusion

Overall then a test ban treaty presented more of a theoretical threat to the UK's acquisition of new nuclear weapons in the period covered by this book; neither Chevaline nor Trident was adversely affected by the prospect of a treaty. Officials of course still worried about the potential impact that such a treaty might have on a new strategic warhead as well as a new tactical warhead and had to plan accordingly. Shortages of fissile material, critical staff shortages, industrial action and old facilities at Aldermaston caused more material delays and threats to the weapons programme than a prospective, or even an actual test ban treaty or testing moratorium. By the time the Trident D5 decision was taken in 1982, the tripartite test ban negotiations had long since been suspended as we saw in Chapter 2; and the future Trident warhead itself would have a planned 30-year service life.[119] AWRE could only work on one new warhead design at a time, so serious work on replacement warhead for the WE177 had to be postponed until the late 1980s and 1990s. It is perhaps ironic that the UK-planned nuclear test codenamed ICECAP, which was due to be fired in late spring 1993, would complete the preparatory trials for the Future Theatre Nuclear Warhead.[120] This never took place as President George Bush reluctantly signed a new law in October 1992 proposed by Congress to ban US testing for a year into law.[121] President Clinton subsequently extended the US moratorium in July 1994, a decision that effectively put paid to any further UK nuclear testing (excluding of course 'activities nor prohibited' by the Treaty). The US moratorium provided a backcloth for multilateral negotiations for a test ban treaty that opened in Geneva in 1994 and which finally produced the text of a Comprehensive Nuclear Test Ban Treaty in 1996.

Notes

1   TNA PREM 15/2038, Future of strategic nuclear deterrent, M/7/2, 1 November 1973, Cabinet Defence Expenditure, Minutes of a Meeting on 30 October 1973.
2   TNA DEFE 72/616, Chevaline and PNR and A3T/ET317 Re-outload, HDP (N) S, 8 July 1980.
3   TNA DEFE 72/616, Chevaline and PNR and A3T/ET317 Re-outload, HDP (N) S, 8 July 1980.
4   TNA DEFE 72/616, Chevaline and PNR and A3T/ET317 Re-outload, HDP (N) S, 8 July 1980.
5   TNA DEFE 24/1344, Annex B, Comprehensive Test Ban (CTB), Military and Security Consequences, Defence Department, 14 January 1977.
6   TNA DEFE 69/1265, Chiefs of Staff Committee, Defence Policy Staff, Review of the Chevaline Project, Report by the Assistant Chief of the Defence Staff (Policy), J. Gingell, Assistant Chief of the Defence Staff (Policy), DP 12/77, Appendix 2 to Annex A to DP 12/77 (Final), 13 July 1977.
7   Kristan Stoddart, *Losing an Empire and Finding a Role, Britain, the USA, NATO and Nuclear Weapons, 1964–70*, Basingstoke, Palgrave Macmillan, 2012, p. 221.
8   John R. Walker, *A History of the United Kingdom's WE177 Nuclear Weapons Programme From Conception to Entry Into Service 1959–1980*, London, BASIC, 2018, p. 18.
9   John R. Walker, *A History of the United Kingdom's WE177 Nuclear Weapons Programme From Conception to Entry Into Service 1959–1980*, London, BASIC, 2018, p. 19.
10 John R. Walker, *A History of the United Kingdom's WE177 Nuclear Weapons Programme From Conception to Entry Into Service 1959–1980*, London, BASIC.
11 TNA DEFE 68/252, Theatre Nuclear Weapons Requirements Committee: WE 177 replacement, 1977 January 1–1979 December 31 Closed file. It is interesting to note that in a briefing prepared in August 1977 for Ministers on US plans for a neutron bomb officials observed that 'Aldermaston has the expertise and ability to design enhanced radiation weapons. The possible development of such a device is one of the options for consideration when the replacement of the existing service tactical weapons comes up for decision'. TNA DEFE 13/2469, S. St John, DS 12 to APS/Secretary of State, Neutron Bomb, 2 August 1977.
12 TNA DEFE 25/433, AUS (OR) to CSA, Theatre Nuclear Weapons – NAST 1231, 8 January 1979.
13 TNA DEFE 24/2116, Hd DS 17 to DUS (P), Successor System Options, 28 January 1980.
14 TNA DEFE 72/302, Theatre Nuclear Weapons Policy Steering Group, Minutes of the Meeting held in Historic Room 25 Main Building on 19 December 1980.
15 TNA DEFE 69/1307, TNWPSG (82) 1, MOD, Theatre Nuclear Weapons Policy Steering Group, Note by the Secretary, 24 August 1982.

16 See, for example, TNA DEFE 69/1307, Nuclear weapons policy: Theatre Nuclear Weapons Policy Steering Group, 1982 January 1–1983 August 31; TNA DEFE 25/678, Tactical nuclear weapons policy steering group (WE177 replacement), 1986 January 31–1987 July 20; TNA DEFE 69/1693, Nuclear anti-submarine capability: replacement for WE177, 1988 September 9–1990 April 24, Closed file; TNA DEFE 71/1210, Equipment Policy Committee (EPC): submission for the future (air launched) theatre nuclear weapon, Staff Requirement SA1244, 1990 October 12–1990 October 12, Closed file.

17 TNA DEFE 23/288, J.M. Legge, Head DS17 to PS/PUS, Trident, Question 12: Saving to be made if the UK were not to renew the existing theatre nuclear capability, 8 January 1981.

18 TNA DEFE 69/1307, TNWPSG (82) 1, MOD, Theatre Nuclear Weapons Policy Steering Group, Note by the Secretary, 24 August 1982.

19 TNA DEFE 23/221, Annex A to RCM/79/1512, Nuclear Costs – Assumptions on Timing, 10 October 1979.

20 TNA DEFE 69/1307, Report on TNWPSG Meeting – 10 June 1983, DUS (P), 11 August 1983; TNA DEFE 72/302, DUS (P) TO DCDS, VCNS et al, Extension in Service of the WE177 B Weapon, 8 October 1982.

21 TNA DEFE 23/221, B. M. Norbury, PS/Secretary of State to DUS (P), The Successor to Polaris, (draft Note by the Secretary of State for Defence to the PM), 31 October 1979.

22 TNA PREM 19/14, Francis Pym to Prime Minister, The Successor to Polaris, 1 November 1979.

23 TNA DEFE 19/181, K. Johnston, AD/DSc 6 to D.S. Bryars, AUS (D Staff), Grey Area Systems, 9 November 1978.

24 TNA PREM 19/4054, Secretary of State for Defence to Prime Minister, UK-Sub-Strategic Nuclear Capability, 1 October 1993.

25 Group Captain W.J. Taylor, 'Engineering on a Nuclear Strike Squadron', *Royal Air Force Historical Society Journal*, Vol. 26, p. 92, Steventon, 2001.

26 TNA DEFE 72/616, Deputy Chief Defence Procurement (Nuclear) to Rear Admiral Sir David Scott, CPE, Polaris A3T – ET317 Warhead, 25 January 1980. The designation for the Chevaline warhead was SF646, but this chapter will refer to Chevaline warheads in the interests of clarity.

27 TNA PREM 19/14, The Nuclear Warhead Test Programme, MOD, 17 May 1979.

28 TNA DEFE 69/1265, Chiefs of Staff Committee, Defence Policy Staff, Review of the Chevaline Project, Report by the Assistant Chief of the Defence Staff (Policy), J. Gingell, Assistant Chief of the Defence Staff (Policy), DP 12/77, Appendix 2 to Annex A to DP 12/77 (Final), 1 3 July 1977.

29 TNA DEFE 69/1265, R.T. Jackling, MOD to E.A.J. Fergusson, FCO, Chevaline, 26 October 1977.

30 TNA DEFE 13/1769, R.T. Jackling, MOD to E.A.J. Fergusson, FCO, Chevaline, 26 October 1977.

31 TNA DEFE 13/1039, Secretary of State for Defence to Prime Minister, Polaris Improvements, 18 September 1975.

32 TNA DEFE 19/240, UK Stockpile Reliability in the Absence of Nuclear Experiments, AWRE, 1 February 1978.

33 TNA DEFE 23/218, Safety and Reliability of the UK Nuclear Weapon Stockpile under a CTB (CTB), Note by DCA (PN), 13 March 1978.

34 TNA DEFE 72/616, KH793 Tactical Production Programme, Appendix 2, A. W. Parsons, DAWP&F, 10 November 1975.

35 TNA DEFE 13/1039, A Report on the Progress and Status of the Chevaline Project, The Main Report, 1 April 1976.

36 TNA DEFE 19/275, John Hunt to Prime Minister, Future of the British Deterrent, Part III – System Options, Annex A, The Present Strategic Force and its Future Life, paragraph 6, 7 December 1978.

37 TNA DEFE 23/222, Ministry of Defence, Chevaline Steering Committee, Minutes of Meeting on Thursday 29 June at 0930, 6 July 1978.
38 TNA DEFE 69/1265, B.M. Day, AUS (OR) to PS/PUS, PS/CDP, SECCOS, Chevaline and Successor Systems, Implications of Arms Control, 14 October 1977.
39 TNA DEFE 13/2469, CSA to PUS, CTBT and the British Nuclear Warhead Programme, 9 August 1977.
40 TNA DEFE 69/1265, Chiefs of Staff Committee, Defence Policy Staff, Review of the Chevaline Project, Report by the Assistant Chief of the Defence Staff (Policy), J. Gingell, Assistant Chief of the Defence Staff (Policy), DP 12/77, Appendix 2 to Annex A to DP 12/77 (Final), 13 July 1977.
41 TNA DEFE 69/742, BATTLER Project Study – Flight Trials Authority, J. Hilton and A.B. Maybury, Final Report, Report No: ST21403, November 1978. A DASO is a series of missile tests fired from a submarine to validate the weapon system and ensure a submarine crew's readiness to use that system.
42 John R. Walker, 'British Nuclear Weapons Stockpiles by Year: 1953–1977', *RUSI Journal*, Vol. 166, No. 4, pp. 10–20, 2020.
43 TNA DEFE 19/208, R.A. Lloyd Jones, Head of DS4 to DC Polaris, DNW, DN Plans, Head of Nat (Coord) (N), Tactical Missiles Stocks for SSBNs, draft submission Future Level of Missile Stocks for SSBNs, 16 October 1974.
44 TNA DEFE 19/208, D.C. Fakley, Head of DSc 6 to Hd DS 4, Tactical Missile Stocks for SSBNs, 24 October 1974.
45 TNA DEFE 19/191, D.C. Fakley, Head of DSc 6, MOD to DAWD, DAWP & F and AWRE, Tactical Outload Requirements for the UK Polaris Force, 21 February 1975.
46 TNA DEFE 19/273, KH 793 Project Status November 1975.
47 TNA DEFE 19/274, Frank Cooper, MOD to L. Plistzky, HM Treasury, Chevaline, 6 July 1976.
48 TNA DEFE 13/1769, R.L.L Facer, MOD to B.G. Cartledge, 10 Downing Street, Chevaline, 29 September 1977.
49 TNA DEFE 72/167, John Elliott, Head of DS2, MOD to J.O. Kerr, Defence Department, Treasury, SALTS Facility – Transfer to ROF Burghfield, 20 June 1979.
50 TNA DEFE 72/523, D.R. Reffell, Rear Admiral, ACNS (P) to VCNS, Introduction of the Trident SSBN Force, DN Plans 21/9/14, 3 November 1980.
51 TNA DEFE 69/742, BATTLER, CWSE Memorandum No. 1779, Chief Weapon System Engineer, 5 January 1979.
52 TNA PREM 65/1265, Draft Minute, Secretary of State for Defence to Prime Minister, Chevaline, 9 January 1979.
53 TNA PREM 65/1265, Draft Minute, Secretary of State for Defence to Prime Minister, Chevaline, 9 January 1979.
54 TNA PREM 19/14, Clive Whitmore, 10 Downing Street to B. Norbury, MOD, Chevaline, 5 November 1979.
55 TNA DEFE 23/220, Clifford Cornford to CNS, 7 June 1979.
56 TNA DEFE 23/220, F.H. East, CWSE to CNS, 3 July 1979.
57 TNA DEFE 23/220, V.H.B. Macklen, DCA (PN) to CDP, Effects of IPCS Dispute on Nuclear Programme and Particularly Chevaline, 5 July 1979.
58 TNA DEFE 23/220, V.H.B. Macklen, DCA (PN) to CDP, Chevaline Warhead Production for Trials and Deployment, 23 July 1979.
59 TNA DEFE 72/616, A. W. Parsons, DAWP&F to CWSE and DCDP (N), Chevaline Production Programmes, 24 September 1979.
60 TNA DEFE 23/221, CNS to Secretary of State, The Chevaline Programme, 19 October 1979.
61 TNA PREM 19/14, Francis Pym to Prime Minister, Chevaline, 1 November 1979.
62 TNA PREM 19/159, Francis Pym to Prime Minister, Chevaline, 18 March 1980.
63 Kate Pyne, 'More Complicated than Expected', in Royal Aeronautical Society, ed., *The History of the UK Strategic Deterrent*, London, Royal Aeronautical Society, 2004.

64 Hansard, HC Deb, Vol. 996 cc204-5W, 17 December 1980.

65 TNA DEFE 72/616, A.W. Parsons, DAWP & F to Capt M.V. Worstall, CPE, MOD, Chevaline Production Aspects, 1 November 1979.

66 TNA DEFE 72/616, A.W. Parsons, DAWP & F to Capt M.V. Worstall, CPE, MOD, Chevaline Production Aspects, 1 November 1979.

67 TNA DEFE 72/616, W.D.S. Scott, Rear Admiral, Chief Polaris Executive to Deputy Chief Defence Procurement (Nuclear), Atomic Weapons Research Establishment, 24 January 1980.

68 TNA DEFE 72/616, D.P. Janisch, HDP (N) S to CPE – Rear Admiral Sir David Scott, Refurbishment of ET 317 Warheads, 29 February 1980.

69 Surveillance rounds were identical to service rounds but contained inert materials instead of fissile materials and were designed to see how the weapon stood up to long-term storage and operational conditions and environments. Such rounds were periodically disassembled to examine for any changes or degradation in the weapon's components that might affect its reliability or safe storage.

70 TNA DEFE 72/616, D.P. Janisch, HDP (N) S to CPE – Rear Admiral Sir David Scott, Refurbishment of ET 317 Warheads, 29 February 1980.

71 TNA DEFE 72/616, W.D.S. Scott Rear Admiral CPE to ACNS (P), Refurbishment of ET317 Warheads, 6 March 1980.

72 TNA DEFE 72/616, Refurbishment of ET317 Warheads, C.C. Fielding, Deputy Chief Defence Procurement (Nuclear)/Director AWRE to Read Admiral J.S. Grove, Chief Polaris Executive, 3 July 1980.

73 TNA DEFE 72/616, HDPN (S), Chevaline PNR AND A3T/ET317 Re-outload, An Assessment by the Defence Procurement (Nuclear) Staff, 8 July 1980.

74 The linkage between Unit 19002 and 'nuclear material' here does suggest that this component was the warhead's secondary (i.e. the thermonuclear part of the warhead containing lithium deuteride), which also included highly enriched uranium.

75 TNA PREM 19/695, Robert Armstrong to Prime Minister, The United Kingdom Strategic Deterrent, MISC 7 (82) 1, 3 March 1982; TNA PREM 19/694, F.H. East, Cabinet Office to R.A. Armstrong, Cabinet Office, Chevaline, 17 February 1982.

76 TNA DEFE 13/1871, Secretary of State for Defence to Prime Minister, Chevaline Acceptance Firings, 30 June 1983.

77 Peter Hennessy and James Jinks, *The Silent Deep: The Royal Navy Submarine Service since 1945*, London, Penguin Books, 2016, p. 467.

78 TNA DEFE 72/616, DGST (N) to CSSE, CPS Co-ord (N), AUS (FS), CPS, C of N, Level of Provision of Polaris A3TK Missiles for Chevaline Weapon System in Service, release of A3T WSAs to meet A3TK REB Production, 30 October 1981.

79 TNA DEFE 19/275, John Hunt to Prime Minister, Future of the British Deterrent, Annex E, International Political Factors, paragraph 8, 7 December 1978.

80 TNA PREM 19/14, The Nuclear Warhead Test Programme, MOD, 17 May 1979.

81 TNA PREM 19/212, R.L.L. Facer, MOD to B. Cartledge, 10 Downing Street, 11 May 1979.

82 TNA PREM 19/14, The Nuclear Warhead Test Programme, MOD, 17 May 1979.

83 TNA PREM 19/159, F. Pym to Prime Minister, British Nuclear Test Programme, 8 January 1980.

84 TNA PREM 19/159, Frank Cooper, MOD to C.A. Whitmore, 10 Downing Street, Prime Minister's Visit to Washington, Successor to Polaris: Timescale, 14 December 1979.

85 TNA DEFE 23/229, The Future of the UK Nuclear Deterrent. A Commentary, DFS (C), C9, Attachment to J.B. Duxbury, Air Commodore, (Secretary Chiefs of Staff Committee) to CDS, CNS, CGS, CAS, The Future of the UK Nuclear Deterrent, 13 August 1979.

86 TNA DEFE 23/221, CSA, MOD to Sir John Hunt, Cabinet Office, The Strategic Deterrent, Future of the UK Nuclear Deterrent, Part III – System Options, Annex E, 12 October 1979.

87 TNA PREM 19/212, R.L.L. Facer, MOD to B. Cartledge, 10 Downing Street, 11 May 1979.

88 See Daniel Salisbury, *Secrecy, Public Relations and the British Nuclear Debate: How the UK Government Learned to Talk About the Bomb, 1979–1983*, London, Routledge, 2020.

89 Ministry of Defence, Defence Open Government Document 80/23, *The Future United Kingdom Strategic Nuclear Deterrent Force*, July 1980, p. 24.

90 TNA DEFE 13/2826/1, Annex A Proposal for an Underground UK Nuclear Tests in 1984, under draft minute from Dr J. A. Davies (Dr) Sc (Nuc)/1 to PS/CSA, British Underground Nuclear Test-1994, May 1992.

91 TNA DEFE 13/2826, M.H. McTaggart, DSc (Nuc) to D (Nuc) R et al., Sunbow Prime 2, Annex A, Proposal for a UK Underground Nuclear Test in 1992, 13 December 1989.

92 The depth of burial was at 1,801 feet. The previous two tests were Kinibito on 5 December 1985 with a depth of burial 1,901 feet and Egmont on 9 December 1984 with a depth of burial at 1,1792 feet. Tests at this depth suggest that yield would have been a yield at about 100 kilotons. *United States Nuclear Tests July 1945 through September 1992*, U.S. Department of Energy, National Nuclear Security Administration, Nevada Field Office, DOE/NV-209-REV, 16 September 2015, https://www.nnss.gov/docs/docs_LibraryPublications/DOE_NV-209_Rev16.pdf. Accessed 3 April 2020.

93 TNA DEFE 23/288, R.M. Hastie-Smith, DUS (P) to PS/SoS, MISC 7-Nuclear Presentation (Costs and the Defence Budget), 23 December 1981.

94 TNA DEFE 72/523, D.R. Reffell, Rear Admiral, ACNS (P) to VCNS, Introduction of the Trident SSBN Force, DN Plans 21/9/14, 3 November 1980.

95 Hansard, House of Commons Debates, Vol. 19 column 985, 11 March 1982.

96 TNA DEFE 13/1871, DUS (P) to PS/|SoS, UK Strategic Deterrent and Arms Control, British Independent Strategic Deterrent and Arms Control, Draft Minute from the SoS Defence and Foreign Secretary to Prime Minister, 26 July 1983.

97 TNA DEFE 25/434, CSA to DUS (P) *et al*, NMWP, Draft Attached Paper The Study of Factors Relating to Further Considerations of the Future of the UK Nuclear Deterrent, 20 September 1979.

98 TNA PREM 19/694, R.A. Armstrong to Prime Minister, The UK Strategic Deterrent, MISC 7 (81) 1, 11 January 1981.

99 TNA PREM 19/694, R.A. Armstrong to Prime Minister, The UK Strategic Deterrent, Background, MISC 7 (81) 1, 23 November 1981.

100 TNA CAB 130/1160, MISC 7 (81), 17 November 1981, Cabinet Nuclear Defence Policy, United Kingdom Strategic Deterrent, Memorandum by the Secretary of State for Defence, Annex C, 17 November 1981.

101 TNA DEFE 23/220, The Future of the UK Deterrent – A Commentary, Attachment 2 to COS 22nd MTG/79, Item 1, 1979.

102 TNA DEFE 13/1871, R.B. Bone, Private Secretary to A.J. Coles, 10 Downing Street, UK Nuclear Deterrent, 4 October 1983.

103 TNA DEFE 23/288, M. Gainsborough, DFA (SS) to PS/PUS, Trident, 8 January 1982.

104 TNA DEFE 13/1871, DUS (P) to PS/SoS, Nuclear Matters, 17 January 1983; TNA DEFE 13/1871, DCDS to CDS, CNS *et al*, Trident, 9 September 1983.

105 TNA DEFE 13/1871, D.J. Fewtrell, Head of DS17 to PS/SoS, UK Nuclear Deterrent, 18 October 1983.

106 TNA CAB 130/1129, Cabinet Nuclear Defence Policy, Future of the United Kingdom Strategic Deterrent: The Present Position, Note by the Secretaries, MISC 7 (80) 1, 29 May 1980; TNA CAB 130/1182, MISC 7 (82) 2, 24 March 1982, Cabinet Nuclear Defence Policy, Special Nuclear Materials for the Defence Programme, Note by the Secretary of State for Defence, 24 March 1982.

107 John R. Walker, 'Potential Proliferation Pointers from the Past: Lessons From the British Nuclear Weapons Program, 1952–69', *The Nonproliferation Review*, 17 February 2012.

108 TNA CAB 130/1182, MISC 7 (82) 2, 24 March 1982, Cabinet Nuclear Defence Policy, Special Nuclear Materials for the Defence Programme, Note by the Secretary of State for Defence, 24 March 1982.

109 TNA DEFE 25/595, John Nott to Prime Minister, Special Nuclear Materials, 31 July 1981.

110 TNA PREM 19/694, Prime Minister Thatcher to President Reagan, 10 September 1981.

111 TNA PREM 19/694, R.L.L. Facer, MOD to B. G. Cartledge, 10 Downing Street, 20 July 1979.

112 TNA CAB 130/1182, MISC 7 9820 3, 25 May 1982, Cabinet Nuclear Defence Policy, Special Nuclear Materials (SNM), Memorandum by the Secretary of State for Defence, 25 May 1982.

113 TNA DEFE 25/595, Note of a Meeting with British Nuclear Fuels held in PUS's Office, 31 March 1982, S. Webb, PS/PUS, 1 April 1982.

114 TNA DEFE 13/1945, Washington telno 237 to FCO, US/UK Cooperation on Special Nuclear Materials (SNM), 27 January 1982.

115 TNA DEFE 13/1945, F.N. Richards, Private Secretary to Clive Whitmore, 10 Downing Street, US/UK Cooperation on Special Nuclear Materials (SNMs), 11 February 1982.

116 TNA PREM 19/685, R.A. Armstrong to C. Whitmore, 14 June 1982. Special Nuclear Materials (SNM) for the Defence Programme, MISC 7 (82) 3, 14 June 1982.

117 Peter Hennessy and James Jinks, *The Silent Deep: The Royal Navy Submarine Service since 1945*, London, Penguin Books, 2016, pp. 514 and 667.

118 TNA CAB 130/1158, MISC 1 (81) 1, Cabinet Official Group on International Aspects of Nuclear Defence Policy, Nuclear Test Ban Policy: Draft Report, Note by the Secretary, 12 January 1981.

119 TNA DEFE 25/1505, Malcolm Rifkind, Secretary of State for Defence to Prime Minister, Comprehensive Test Ban – Activities Not Prohibited, 16 February 1995, p. 3.

120 TNA DEFE 13/2826-1, J.A. Davies (Dr) Sc (Nuc)/1 to PS/CSA, British Underground Nuclear Test – 1994, Draft Minute from CSA to Secretary of State, May 1992.

121 Frank N. von Hippel, 'The Decision to end U.S. Nuclear Testing', *Arms Control Today*, Vol. 49, December 2019.

6   Conclusion

British Nuclear Weapons and the Test Ban 1974–1982

The central question at the heart of this book was whether the UK managed to pursue two ostensibly competing objectives simultaneously between 1974 and 1982: a comprehensive nuclear test ban treaty designed to prevent the development of nuclear weapons and their proliferation to non-weapon states whilst trying to sustain and modernise the UK's own nuclear weapons capabilities. Given the level of knowledge and expertise available to the UK in the late 1970s, and indeed to the US, and the way in which British nuclear weapons were designed at that time (to very fine engineering tolerances), the pursuit of this objective was simply not achievable. A truly comprehensive test ban treaty of indefinite duration proved to be incompatible with maintaining nuclear weapons stockpile reliability and safety; furthermore, a treaty would foreclose options to design and deploy new strategic and tactical warheads for the UK nuclear arsenal unless these could be designed and tested before a ban entered into force. The decision to opt for a three-year test ban treaty instead of even a five-year treaty was driven by US and UK fears over stockpile reliability and safety; and if testing resumed after the expiration of such a treaty, that would then enable testing of new warheads to replace Chevaline and the WE177. However, such a treaty, had it ever been agreed, would have had virtually zero value for non-proliferation as the key non-weapons states along with France and China would see no security or political benefit whatsoever for them. The election of Margaret Thatcher in May 1979, who in her own words was 'very suspicious' and 'worried' about a treaty, made it more likely than not that the UK would not have signed such a treaty unless there were clear guarantees on stockpile reliability and safety along with an ability to test new designs for the Polaris replacement system – Trident. That the UK was not faced with such a stark choice – sign or reject a treaty – was largely down to the impasse in the tripartite negotiations over the scope and content of the treaty's verification regime, including the row over the number of National Seismic Stations (NSS) on UK territory, the inability to resolve the 'permitted experiments' conundrum and the decision of the Reagan Administration to review US policy on the desirability or otherwise of a test ban treaty for US national security. Whilst the US deliberated, the tripartite negotiations at US request were suspended sine die; the final nail in the treaty's coffin was finally driven in by the Reagan

DOI: 10.4324/9781003375708-6

Administration's announcement in early 1982 that a CTBT was no longer in US interests. The idea of a CTBT, however, would be resurrected by President Clinton in the 1990s, and a treaty was finally negotiated between February 1994 and August 1996 in the Conference on Disarmament. As a prelude to this process, the Clinton Administration announced a moratorium on further nuclear testing on 3 October 1992, which ironically enough put paid to plans to conduct a UK nuclear test codenamed ICECAP, which was only a few weeks away from being fired at the time of the announcement. As the UK had nowhere else to test, this decision effectively put paid to British nuclear testing for the foreseeable future. The stark consequences of direct dependence on the US for access to nuclear testing facilities had finally come home to roost. Indeed officials had recognised this stark reality back in 1978 when the Prime Minister was informed that

> whilst we are participating in negotiations on a CTBT, this must inevitably be very much as the junior partner since we are entirely dependent on the US for test facilities and in the final analysis must be governed by their decisions.[1]

Though even if the UK had access to its own test site, it seems likely that the global and domestic pressures for an end to nuclear testing would have been too great for any British government to resist.

Arguments over verification, primarily on-site inspection, had scuppered the 1958–1962 Conference on the Discontinuance of Nuclear Weapons Tests. The Soviet Union refused to accept on-site inspections in the treaty – something that both the UK and the US regarded as essential for an effective treaty as there would be far too many seismic events in Soviet territory that would look suspicious and as such might be clandestine nuclear tests given their location and depth. By the 1970s, the Russians were still objecting to verification, though this time they were prepared to accept the US demand for ten NSS on Soviet territory along with some form of on-site inspection as long as it was not mandatory. However, there was a sting in the tail – the UK would have to accept ten NSS as well, both in the UK and on its dependent territories. Until such times that the UK accepted this counter-proposal, the Russians refused to discuss any of the technical details that would need to be addressed on the practicalities and implementation of the NSS system. The UK, after much careful thought, stuck with its proposition that ten stations in the UK could not be justified technically, which was undoubtedly true, and would only accept one station at the existing seismic station site at Eskdalemuir. This impasse continued for over a year and the new government of Margaret Thatcher agreed with the position of its Labour predecessor. Although this was not by any means the only issue holding up negotiations of a treaty, Review Conference language and the role of permitted experiments being two other significant obstacles to progress, the US put pressure on the UK to show some willing and compromise on the numbers. Washington felt that the UK should be prepared to agree to six NSS;

and certainly, after June 1979, some US officials started to wonder whether this was a convenient excuse for the new Thatcher government to avoid committing to a treaty. Thatcher's critical comments on the Treaty to Cyrus Vance in June 1979 in London were widely disseminated around the US official community. However, there is no evidence whatsoever that the UK regarded the NSS question as an escape clause, or indeed as a convenient cover for a private change in its national public position in support of a treaty; on the contrary, the position was carefully reviewed and debated at length and great detail by FCO, MOD and Cabinet Office officials as well as Ministers.[2] There is no sign of bad faith, or any duplicity in minutes and papers at the Cabinet Official Group on the International Aspects of Nuclear Defence Policy, the Ministerial committee GEN 63 and the later MISC 7 committee that addressed nuclear defence issues in the Thatcher government. As was the British civil service way of doing things, negotiating objectives and instructions for the delegation were carefully debated and formulated, especially when it touched on nuclear defence matters and ways had to be found for squaring some very challenging circles. Although the number of officials and ministers involved in such high-level nuclear policy-making was extremely small, that does not mean the issues were treated lightly or cynically. Cabinet Office, Foreign Office and MOD officials worked out the details and recommendations for senior Ministers and the Prime Minister with the pros and cons of the various policy options carefully weighed and assessed. Whilst it is clear from the archive at Kew that FCO officials, and certainly the Ambassador in Geneva John Edmonds, argued the case for a CTBT robustly in foreign policy terms, they were not blind to the security dimensions and concerns as expressed by their MOD and AWRE colleagues. As Edmonds himself noted, the FCO and MOD worked well together on the test ban, though by the end of 1980 we start to see clearer divergences between the FCO and MOD over the future direction of UK testing policy.[3] In short, foreign policy is rarely a matter of clear-cut choices and the complex and interlinked pros and cons in play during the 1977–1980 CTBT negotiations serve as a powerful reminder of that reality. In the nuclear weapons context, expertise resides in an extremely small group of experts. As such it is almost impossible for non-experts such as Ministers and FCO officials to gainsay or contradict the advice that comes from such an exclusive epistemic community. This reality appears repeatedly in the story of the tripartite negotiations.

Negotiating with the Americans was never easy. The multitude of agencies, the size of their staffs in Washington and beyond in the weapons laboratories and the propensity of the interagency to leak information to the press all added to the problem facing the UK during the tripartite test ban negotiations. Securing agreement in Washington, especially where the policy is highly contested – as it was quite clearly on the CTBT – is challenging to say the least. Delays in issuing instructions to the US delegation in Geneva caused by protracted arguments over what position to adopt were common place. And as had been the case in the 1950s and early 1960s, there was a strong animus against the very idea of a CTBT in powerful sectors of the Washington inter-agency process.

However, since the UK was so beholden to the US for support for its nuclear weapons programme through the 1958 Mutual Defence Agreement and 1963 Polaris Sales Agreement, it could hardly not work closely with the US on any international negotiation touching on nuclear defence such as those for a test ban treaty. In fact, it would seem that during the negotiations the UK spent far more time in bilateral consultations with the US than in formal trilateral settings with the USSR.[4] Regardless of the nuclear dimension, there would have been a natural predilection in the FCO, MOD and Downing Street to keep close to the US and discuss and agree on common negotiating positions and objectives. In this case, once the US finally woke up, partly at British urging, to the risk a treaty could pose for nuclear weapons stockpile reliability and safety questions, the eventual US decision to opt for a three-year limited treaty inevitably chained the UK to that outcome too. Indeed Dr David Owen, the Labour Foreign Secretary, felt aggrieved that the UK weapons community had fired up their US counterparts with anxieties over stockpile reliability issues. However, it is more likely that the initial US policy on a test ban had not been fully agreed around the agencies. The US weapons labs and the JCS would not have needed any British cajoling on this matter. It is also clear that UK concerns over safety and reliability were not new, or were only newly discovered during the negotiations, they had been clearly articulated more than a year *before* the negotiations began.[5] And despite, the 'special relationship' the US was not always ready to take into account British views – even when pressed by no less a personage than the Prime Minister himself. But then again there were occasions when the US was very interested in the UK experience from its self-imposed testing moratorium in the 1960s, early 1970s, which is a reminder that the flow of warhead and nuclear information was not all one way. There were benefits to the US too.

When the UK stated that it would only accept one NSS, this posed a threat to the US requirement that there would need to be ten NSS on Soviet territory in order to ensure that the test ban treaty was verifiable and as such stand a chance that the Senate would give its advice and consent to the treaty's ratification. Consequently, Washington put some pressure on the UK to compromise as the Carter Administration considered NSS as the major stumbling block to agreeing a treaty, interestingly enough the stockpile safety and reliability problem did not seem to be the higher priority. London did not share that assessment and was more than slightly peeved by US pressure on this question. The problem really only went away when the tripartite negotiations were suspended at the request of the new Reagan Administration, but it would not be fair to say that it was the UK that was wilfully obstructing the negotiations. US positions, not to mention Soviet obduracy and insistence on technically unsound and illogical proposals, were arguably much greater barriers to preventing agreement from being reached. As we have seen, Margaret Thatcher was hardly a staunch supporter of a CTBT – quite the contrary, but she could not afford to antagonise President Carter given his very personal support for a treaty since the UK would shortly have to ask for the transfer of Trident C4 SLBMs to replace Polaris as there was no viable alternative if the UK wanted to remain in the nuclear business. She therefore muted her objections. UK nuclear dependency on the US, particularly at such a critical juncture, inevitably constrained the UK room for manoeuvre, but the spat over the number of NSS does not appear to have had, or threatened to have had, any adverse impact on a UK request for Trident. The bigger issue for President Carter was ensuring that SALT II was ratified by the Senate in 1979, so he was keen that the UK should delay asking him formally for Trident until that process had been completed. Carter's concern was that a British request would complicate matters by raising such a contentious issue – namely the transfer of MIRV technology – with the Senate. In any case, under Thatcher, the UK now put a CTBT as a long-term objective for its arms control policy, which was more or less the same as the US position under Reagan. However, the UK could not, and did not, abandon a CTBT completely, hence the reason for it being a long-term objective. A CTBT was thus off the immediate agenda and thus no threat before the vital Trident warhead proving tests were required in the mid-1980s

In retrospect, the prospects for a truly comprehensive nuclear test ban treaty as an outcome of the tripartite negotiations were always slim, and the chances of securing agreement on even a short-duration treaty of three years proved similarly remote. Once again, a comprehensive test ban treaty had proved unattainable. Ultimately the requirements of the UK nuclear weapons programme took precedence and the Reagan Administration's policy on a test ban certainly helped by taking the spotlight off the UK. In any case, as we have seen, the shortage of critical craftsmen at AWRE and ROF Burghfield and industrial action by the trade unions posed a greater and more immediate threat to the UK weapons programme's objectives and timelines. As Victor Macklen, the key official in the UK nuclear programme from the 1950s, observed in June 1974,

> British Governments have delayed nuclear weapon programme decisions, played with unilateral gestures, pursued disarmament measures or proposals. All Governments have had a conscience over such weapons, but all have despite uncertainty and hesitation in the end backed the British nuclear defence capability.[6]

Macklen's comment also very clearly applies to the period 1974 to 1982, though Margaret Thatcher evidently did not have a conscience over such weapons in the way implied by Macklen given her firm belief in the fundamental importance of the nuclear deterrent for UK security. So in the end the UK failed to square the circle by securing a treaty that would prevent qualitative improvements in nuclear weapons design and constrain further horizontal proliferation to the likes of Pakistan whilst simultaneously maintaining and modernising its own nuclear weapons programme – strategic and tactical. Neither the political will for a treaty nor the scientific and technological basis for maintaining nuclear weapons under an indefinite effectively verifiable CTBT existed between 1977 and 1982. For that, we would have to wait until the mid-1990s, which opened a new chapter in the history of the comprehensive nuclear test ban treaty and

British nuclear weapons. By then the political and technical landscape had rather fundamentally changed with the end of the Cold War and the existence of far greater capabilities to sustain a nuclear weapons stockpile in a world without testing. How all this proved possible will, hopefully, be discussed one day in a future book.

Notes

1 TNA DEFE 19/275, John Hunt to Prime Minister, Future of the British Deterrent, Factors Relating to Further Consideration of the Future of the United Kingdom Nuclear Deterrent, Part I, The Politico-Military Requirement, paragraph 33, 7 December 1978.
2 See, for example, TNA 66/1532 and 1533 Comprehensive Test Ban (CTB): Policy Review 1981; TNA FCO 66/1586 to 1591, Comprehensive Test Ban (CTB): Policy 1982.
3 TNA FCO 66/1473, A. Reeve, Arms Control and Disarmament Department to Mr P.H. Moberly, MISC 1 Meeting on CTB, 20 November at 3. 00 pm, 19 November 1980.
4 See, for example, the wide and detailed UK-US technical consultations that took place in 1977 alone, which are recorded in TNA FCO 66/887, UK/US Bilaterals 2nd Round, 1977.
5 Denis Fakley, Statement at an informal meeting of the CCD, CCD/492, pp. 1–5, 21 April 1976.
6 TNA DEFE 19/187, V.H.B. Macklen, DCA (PN) to Mr Omand, APS/S of S, British Nuclear Policy 1945–1952, 19 June 1974. Macklen was commenting on the draft of Margaret Gowing's official history of the UK nuclear deterrent, *Independence and Deterrence: Britain and Atomic Energy 1945–1952*. I am grateful to Professor Matthew Jones for drawing my attention to this document.

Postscript

With the prospects for a new set of CTBT negotiations looking rosier at the end of 1992 following the election of a new US President, the Defence and Foreign Secretaries jointly minuted the Prime Minister setting out recommendations on how the UK should react.[1] As we saw between 1977 and 1980, and as Malcolm Rifkind and Douglas Hurd again made clear, it was the 'considered judgment of those responsible for the UK nuclear programme that continued testing is necessary to ensure the safety of the existing stockpile could be underwritten as the weapons age and are modified over time'.[2] From the start of the CTBT negotiations two years later in February 1994, the UK believed that experiments, involving some, relatively low, nuclear yield (a 100 tons ceiling was the preferred UK option at that time) would be needed to allow the continued underwriting of the safety and performance of existing and future nuclear weapons.[3] Unfortunately for the UK, the Americans now preferred only a few pounds of nuclear yield for activities not prohibited – a position much changed since the tripartite negotiations where the US preferred figure was 100 lb as we saw in Chapter 4.[4] Aldermaston, however, had assessed that a limit of 4 pounds was the lowest limit for single-point safety tests, but with no margin for error. Experiments at that level would only contribute to one area of weapons physics.[5] Furthermore, new designs of modern thermonuclear weapons could not confidently, or responsibly, be brought into operational service without testing.[6] A key word here is 'responsibly' as that implies that safety was a key consideration. MOD officials noted that the limitations that a CTBT would impose on future British nuclear options were unwelcome – a classic civil service understatement, but in entering negotiations in good faith the UK had accepted that a degree of constraint on its freedom was a price worth paying for the benefits of a CTBT. These were primarily to constrain the ability of would-be proliferators to develop and prove sophisticated nuclear weapons, which provides an echo of the concerns in the late 1970s over proliferation and in particular Pakistan's nuclear ambitions.[7] In addition, the outcome of the NPT Extension Conference scheduled for April–May 1995, where the UK was seeking an indefinite extension of the 25-year treaty, could well hinge on the attitudes and policies of the nuclear weapons states to arms control and disarmament. Non-aligned non-nuclear weapons states attached high importance on the UK, US, Russian Federation,

France and China fulfilling their Article VI obligations, 'to pursue negotiations in good faith on effective measures relating to cessation of the nuclear arms race at an early date and to nuclear disarmament'. A CTBT negotiation underway and likely to succeed would be crucial in such a context. Much of this sounds familiar to the issues, especially the relationship between nuclear defence and arms control, discussed in Chapters 2 and 4. Core arguments and issues do not seem to have really changed at all, including the consequences of the straitjacket imposed on the UK by its dependence on the US for underground nuclear testing facilities and the wider nuclear collaboration under the 1958 MDA, coupled too with the vagaries, swings and roundabouts of the US interagency process and domestic politics. However, there was one big difference – there was a clear determination in the White House that this time round there would be a CTBT, and that clearly helped as did the ending of the Cold War. President Clinton would succeed where Eisenhower, Kennedy and Carter had failed. In doing so, he squared Britain's nuclear defence and arms control circle.

Notes

1 TNA PREM 19/4054, Douglas Hurd and Malcolm Rifkind to Prime Minister, Nuclear Testing, 14 December 1992.
2 TNA PREM 19/4054, Douglas Hurd and Malcolm Rifkind to Prime Minister, Nuclear Testing, 14 December 1992.
3 TNA DEFE 25/1015, D.B. Omand to PS/Secretary of State, Comprehensive Test Ban – Activities Not Prohibited, 2 February 1995.
4 'Activities not prohibited' was the new name for 'permitted experiments', which had been the term of art in the 1977–1980 tripartite negotiations.
5 TNA DEFE 25/1015, D.B. Omand to PS/Secretary of State, Comprehensive Test Ban – Activities Not Prohibited, Annex, Implications of ANP levels on capability maintenance, 2 February 1995. Single-point safety tests are designed to ensure that any accidental detonation at any single point of the high explosive used in a nuclear weapon's implosion system would not lead to a full nuclear yield.
6 TNA PREM 19/4054, C.N.R. Prentice, FCO to J.S. Wall, Nuclear Testing, 10 Downing Street, 25 September 1992.
7 TNA DEFE 25/1015, D.B. Omand to PS/Secretary of State, Comprehensive Test Ban – Activities Not Prohibited, paper attached to the submission, 2 February 1995.

Annex

UK Nuclear Tests Nevada Test Site May 1974–June 1986

NO.	CODE NAME	YIELD KILOTONS[1]	DEPTH OF BURIAL[2]	US LAB	DATE	PURPOSE	COMMENT/CONTEXT
1	**FALLON**	35	1,530 ft	LLNL	23 May 1974	Hardened warhead for Chevaline to protect against neutrons generated by ABM explosion.[3] A device test was held to prove the 'Harriet' warhead design of warhead for the KH 793 project. The results were satisfactory.[4] The test 'successfully met its prime objectives'.[5] Tests necessary for miniaturised hardened warhead being developed to preserve the range of the new Chevaline front end.[6] The planned yield was 35 kilotons.[7] 'low-intermediate yield' (20–200 kilotons)[8]	Drill back release of radioactivity at site only.
2	**BANON**	20–150	1,761 ft	LLNL	26 August 1976	'The test now planned was to improve the effectiveness of our existing system'.[9] The Fallon test of a radically new design had not given sufficient information to be able to guarantee to keep it in a reliable and safe condition over the expected 10–15 years of service life. To acquire this information a further one or two tests of basic design were needed.	Drill back release of radioactivity at site only. The Defence Secretary, Roy Mason, noted in a letter to Labour Party General Secretary Ron Hayward on 1 December 1975 that the test planned for 1974 was not concerned with the development of a MIRV warhead.[13]

						As a bonus these tests might also enable the UK to economise on the manufacture of the warhead, especially in its use of plutonium.[10] This test was intended to assist in a fundamental re-examination of whole system weight/range, to clarify 'exchange ratio', and as a prototype warhead to test against future defensive ABM. These elements were further examined at the Cresset (Fondutta) test in April 1978.[11] 'Low intermediate yield 20–150 kilotons'[12]	Mason also informed the Prime Minister that the yield of UK tests would be 'well below the threshold of 150 kilotons agreed bilaterally by the US and USSR at the Summit Meeting in Moscow in July 1974'.[14] Mason also noted that 'to fit in with the US schedule of tests, long term planning for our own possible requirements has to be carried out on a contingency basis, sometimes even before it is certain that a test will be necessary'.
3	**FONDUTTA**	100±14	LLNL	2,077 ft	11 April 1978	To prove the performance of a 'reduced weight version of the Chevaline warhead'.[15] And to improve AWRE's knowledge of the Chevaline design.[16] It also included 'a significant design change'.[17] 'test will be of an improved Polaris warhead'.[18]	The original plan had been to conduct this test in October 1978, but following Cyrus Vance's visit to Moscow and the possibility of test ban talks, the MOD saw advantage in bringing it forward to March 1978.[26]

(*Continued*)

NO.	CODE NAME	YIELD KILOTONS[1]	DEPTH OF BURIAL[2]	US LAB	DATE	PURPOSE	COMMENT/CONTEXT
						'The purpose of this test was to improve knowledge of warhead design and was related to the existing programme for maintaining the effectiveness of the Polaris force'.[19]

A new light-weight warhead for Chevaline was tested – Cresset (Fondutta), using the Findhorn/Firstrate primary and part of the Battler modification to the Chevaline payload, work on which stopped in January 1979.[20] The test 'will be of an improved Polaris warhead'.[21]

Findhorn was the UK code for the device tested at Fondutta.[22] A lighter warhead would extend the range of the Polaris missiles, provide more sea room and thus greater insurance against improvement in Soviet Anti-Submarine Warfare (ASW) capability.[23]

'My technical experts and their US colleagues are however confident that this successful test demonstrated our capability, should we choose, to put a lightweight warhead into service in later Chevaline missiles giving an increased range for the missile and therefore an increase in the sea room available for our Polaris submarines'.[24] | |

4	QUARGEL	47±5[27]	1,178 ft	LLNL	18 November 1978	'The yield of the device was about 100 kilotons with a margin of error of 14 kilotons either way compared to an expected yield of 106 kilotons'.[25]	Drill back release of radioactivity at site only. The UK codename for the device and test was GAVEL.[37]
						To 'largely repeat' the technology tested at Fondutta.[28] It would 'explore further the new warhead principles' of Fondutta.[29] Quargel was 'a lower yield device based on the same design principle'.[30]	In the absence of a CTBT, DCA (PN) had in a mind a test frequency beyond 1979 of about one per year, or perhaps two in every three years.[38]
						The Quargel device was physically smaller than the one tested at Fondutta.[31] This reduced size meant that the design might be suitable for a pointed Re-entry Body (RB), rather than the blunt-nosed slower Chevaline B. This is our first attempt at a device which would allow a high-speed re-entry vehicle for a ballistic missile.[32]	The test had been delayed from June to December at the PM's request in June.[39] The US asked if the test could be brought forward to mid-November.
						'As expected, the predicted yield was 52 ±5 kt. The most reliable immediate yield measurement gives a figure of 47±5 kt. It been noted in the planning stages that the maximum credible yield would be 75 kilotons with an expected yield of 63 kilotons.[33] This is our first attempt at a device that would allow a high-speed RV for a ballistic missile.	

(Continued)

NO.	CODE NAME	YIELD KILOTONS[1]	DEPTH OF BURIAL[2]	US LAB	DATE	PURPOSE	COMMENT/CONTEXT
						The results are encouraging and comparable with the results from the Poseidon design. However, before we could bring such an experimental device into safe and reliable service, we would need the information from the next test for which we are currently seeking permission to fire in July 1979. If this test were successful it would begin to open the possibilities of increasing the yields of these devices and would certainly open the door to far more exchange with the Americans on their devices of a similar nature'.[34] 'and which, if successful would provide much needed background information for reliability'.[35] 'We are also very keen to get the Quargel test under our belt because it should put us in a position to respond to the full requirement for the warhead for any successor system that might be chosen'.[36]	'My experts cannot guarantee that a need for a British nuclear test for stockpile maintenance purposes will not arise and in that case existence of an already drilled hole could save six months in preparation time for what might be essential safety or reliability tests'.[40] This is the first and seemingly only sign in the archives that the UK might need to conduct such tests.

| 5 | NESSEL | 35–45 | 1,522 ft | LLNL | 29 August 1979 | The maximum credible yield for containment purposes was set at 55 kilotons. The expected yield was between 35 and 45 kilotons.[41]

This was for a much more advanced device than Quargel.[42] It was a 'new design concept'. Nessel was of 'exceptional technical importance' for future nuclear weapons. It was assumed that 'future warheads would be constrained in size and weight'.[43] The novel feature of the design was the 'very small nuclear trigger'.[44] This development would 'greatly broaden our capability to design small "packagable" nuclear warheads which could be applied to a wide range of theatre or strategic delivery systems'.[45] The Nessel test would 'begin to open the possibilities for increasing the yields' of warhead designs suitable for high speed re-entry vehicles.[46] After the Thatcher government came to power, the test was described as related to 'proving a British warhead design for a successor system'.[47] 'The aim of the two most recent UK tests and that aimed for August was to provide information on very small and hard warhead designs suitable for MIRVed systems, but with a nuclear yield lower than that required'.[48] | Gas sampling release of radioactivity detected at site only. DICEL was the UK codeword for the test device and test.[54] Nessel was the US codeword.

This test would also be 'the first in which we can incorporate new and sophisticated diagnostic techniques which will greatly enhance its long-term value. In particular, the new measurements will help designers to interpret the sort of laboratory experiments which will provide the only new design data available during a CTBT'.[55]

The diagnostic techniques focused on the performance of the primary: 'If successful the results from these advanced diagnostics will give my experts a clearer picture in the crucial stages of the implosion of the very small trigger which lights the thermonuclear components'.[56] |

(Continued)

NO.	CODE NAME	YIELD KILOTONS[1]	DEPTH OF BURIAL[2]	US LAB	DATE	PURPOSE	COMMENT/CONTEXT
						It was also noted that if this test was successful, 'it would then be possible to produce a hard light weight UK warhead of [deleted in original] suitable for an SLBM carrying MIRVs'.[49] 'The August test is particularly important for the Chevaline programme'.[50] "The initial results were described as "satisfactory"'.[51] 'the expectations of our weapon designers were fulfilled and may be exceeded. The yield of the shot, as recorded seismically, appears to be in accordance with prediction and the main components of the test device appear to have operated very well'. 'and gives confidence both in our design ability and in the prospects for developing a warhead to meet a future requirement for a successor deterrent system'.[52] Both LANL and LLNL were very interested in this UK nuclear test and 'were very keen to see it conducted'.[53]	The July 1979 test (Note: delayed until August) will be an especially significant test for our weapons designers as the US have agreed to carry out the most extensive diagnostic measurements, including some parameters we have never been able to measure before. Moreover, both US weapon laboratories are now very interested in this particular UK test and are keen to see it done.[57]
6	**COLWICK**	20–150	2,077 ft	LLNL	26 April 1980	'An experimental device which might almost match the current US performance has been designed and there are provisional plans to test this at Nevada in mid-1980'.[58]	Gas sampling release of radioactivity detected at site only. This was probably about 100 kilotons given the DOB.

| 7 | **DUTCHESS** | < 20 | LANL | 1,400 ft | 24 October 1980 | To test a mechanism for varying the yield of future tactical nuclear weapons.[64] Probable code name Lute Prime.[65] | 'preliminary indications are that the test device achieved its designed nuclear yield . . . The full records of the test will be available shortly and the subsequent detailed analysis will allow us to judge how well the device operated and how far we have met our objectives. The initial reactions are all favourable'.[59] 'A further successful test in mid-1980 should enable a full yield requirement for a successor system warhead to be met'.[60]

This was the first in a series of five tests between 1980 and 1981 that the PM had approved on 11 January 1980 following a proposal from the MOD dated 17 December 1979.[61] | The UK codeword for the device and test was DINGBAT.[62]

The Defence Secretary informed the Prime Minister in July 1979 that, 'Our warhead development programme aimed at successor systems and modernisation of our tactical nuclear weapons, would require at least two more successful tests before a (test) ban comes into effect. Making allowance for a possible test failure means that we should ask US authorities to seek Presidential authority for at least three further British nuclear tests in 1980'.[63]

The planned warhead development for Trident and theatre nuclear weapons required a threshold test ban of not less than about 30 kilotons.[66] |

(Continued)

NO.	CODE NAME	YIELD KILOTONS[1]	DEPTH OF BURIAL[2]	US LAB	DATE	PURPOSE	COMMENT/CONTEXT
8	**SERPA**	20–150	1,879 ft	LLNL	17 December 1980	'to evaluate a possible warhead for the system to replace Polaris'.[67] Device known as Hurdle Prime.[68] 'preliminary results indicate that the device functioned successfully'.[69]	–
9	**ROUSANNE**	20–150	1,697 ft	LANL	12 November 1981	'The yield switching mechanism was shown to be satisfactory . . . and we obtained some new information to widen our technology base. The test was important because the device incorporated some new warhead features that are beyond our present powers to evaluate quantitatively. Investigation of the behaviour of such features is still possible only by underground testing'.[70] 'all the objectives of the test were successfully met'.[71] 'an important part of our progress towards designing warheads for our new Trident missiles in the 1990s'.[72] 'aimed at investigating further the feasibility of mechanisms for switching the nuclear yield of warheads for theatre applications'.[73]	Dingbat II seems to have been the device.[74] The test device for Rousanne was code-named Lute.[75]

10	**GIBNE**	20–150	1,869 ft	LLNL	25 April 1982	'will inform an important part of our progress towards designing warheads four our new Trident missiles in the 1990s'.[76] 'preliminary indications are that the operation went according to plan and that the test was successful'.[77]	Drill back release of radioactivity at site only. This was possibly the HURDLE PRIME device.[78] The original plan had been to hold this test in December 1981, and MOD would have preferred to hold it earlier.[79]
11	**ARMADA**	>20	869 ft	LLNL	22 April 1983	First Holbrook test.[80]	Drill back release of radioactivity at site only.
12	**MUNDO**	20–150	1,857 ft	LANL	1 May 1984	–	–
13	**EGMONT**	20–150	1,792 ft	LLNL	9 December 1984	Trident validation test.	–
14	**KINIBITO**	20–150	1,901 ft	LANL	5 December 1985	Trident validation test.	–
15	**DARWIN**	20–150	1,801 ft	LLNL	25 June 1986	Final Trident validation test – Holbrook III.[81]	–

Notes

1 Experience had shown that the actual yield of a nuclear test can exceed the design yield by a significant margin – up to 50% for the primary fission component of a device. US policy under the Threshold Test Ban Treaty (signed July 1974) was deliberately conservative: design yields were limited to 30% less than the Treaty threshold of 150 kilotons. The National Archives (TNA) CAN 130/1158, MISC 1 (81) 3, Cabinet Official Group on International Aspects of Nuclear Defence Policy, Nuclear Test Ban Policy: Report, Note by the Secretary, 6 February 1981, Annex D Technical Aspects of a Low Threshold Test Ban. This means that all UK tests after Fallon listed here would not have had yields greater than about 105 kilotons. UK devices were transported to Nevada in component form and only assembled by AWRE on-site prior to them being handed over the US to lower down the borehole, which had been drilled by the US.

2 US tests of less than 20 kilotons were conducted in shafts ranging from about 600 to about 1,400 feet, which suggests that most UK tests were in the tens of kilotons given their depth of burial (DOB). The exception is Dutchess, which is the one test listed at less than 20 kilotons.

3 TNA DEFE 19/207, The Project Definition Study of Super Antelope (KH 793) Report of the KH 793 Project Review Board, 3 October 1972.

4 TNA DEFE 19/273, KH 793 Project Status, November 1975, F.H. Panton, ACSA (N) to CSA, 8 December 1975.

5 Hansard, Mr Mason, Oral Answers, Column 177, 14 January 1975.

6 Kristan Stoddard, *The Sword and the Shield: Britain, America, NATO and Nuclear Weapons, 1970–1976*, London, Palgrave Macmillan, 2014, p. 162.

7 National Security Archive, George Washington University, Briefing Book #: 770 edited William Burr, 16 July 2021, Document 18, https://nsarchive.gwu.edu/document/23810-memorandum-conversation-nuclear-release-agreement-labour-government-s-defense-review, Memorandum of Conversation between Sir John Hunt, Cabinet Secretary et al. and Dr Henry Kissinger et al. Secretary of State and National Security Adviser, 26 April 1974.

8 TNA FCO 46/1829, Public Relations Plan for the Fondutta/Findhorn Nuclear Test, Attachment C, April 1978. The Threshold Test Ban Treaty was signed in July 1974, though it did not enter into force until December 1990. However, the US (and the UK) observed the limit of 150 kilotons until then, and of course thereafter.

9 TNA CAB 128/59/22, Cabinet, CM (76) 22nd Conclusions Agenda Item 2, Nuclear Test, 3 August 1976.

10 TNA PREM 16/1181, Roy Mason to Harold Wilson, Polaris Improvement Programme – Nuclear Testing, 28 May 1975.

11 Kristan Stoddard, *The Sword and the Shield: Britain, America, NATO and Nuclear Weapons, 1970–1976*, London, Palgrave Macmillan, 2014, p. 164.

12 TNA FCO 46/1829, Public Relations Plan for the Fondutta/Findhorn Nuclear Test, Attachment C, April 1978.

13 TNA DEFE 13/1247, Roy Mason to Ron Hayward, 1 December 1975.

14 TNA DEFE 13/1247, Roy Mason to Prime Minister, Nuclear Tests, 29 October 1975.

15 TNA DEFE 13/1478, British Nuclear Test Programme, V.H.B. Macklen, 19 April 1978.

16 TNA DEFE 13/1477, British Nuclear Test Planned for March 1978, V.H.B. Macklen, 17 November 1977. The Fondutta test was postponed from March to April.

17 TNA DEFE 13/1478, British Nuclear Test Programme, 27 April 1978 and TNA DEFE 13/1478, Worded as '*a significant advance in design*' in Macklen's draft, 19 April 1978.

18 TNA DEFE 13/1769, Fred Mulley to Prime Minister, British Nuclear Test – March 1978, 20 July 1977.

19 TNA CAB 128/63/10, Cabinet, CM (78) 10th Conclusions Agenda Item 3, 16 March 1978.

20 TNA DEFE 69/742, BATTLER, CWSE Memorandum No. 1779, Chief Weapon System Engineer, 5 January 1979.

21 TNA FCO 46/1566, Fred Mulley to Prime Minister, British Nuclear Test – March 1978, 20 July 1977.

22 TNA DEFE 24/1361, AWRE, Aldermaston Classification Notice No.43, Codenames of Underground Tests of UK Nuclear Devices at NTS, S.K. Hutchison, 7 March 1980.

23 TNA PREM 16/1564, John Hunt to Prime Minister, Military Nuclear Issues, 25 October 1977.

24 TNA FCO 46/1830, Fred Mulley to Prime Minister, British Nuclear Test – 11 April, 27 April 1978.

25 TNA FCO 66/1082, Fred Mulley to Prime Minister, British Nuclear Tests – 11 April, 27 April 1978.

26 TNA FCO 66/897, D.C. Fakley, Director, DSc 6, MOD to J.C. Edmonds, ACDD, FCO, UK Programme for Maintaining the Effectiveness of the Deterrent, 20 April 1977.

27 TNA DEFE 25/335, British Nuclear Test, V.H.B. Macklen, 20 November 1978. The planned yield was 52 kilotons and the actual yield 47 kilotons. Both yield figures are +/– 5 kilotons. These figures were released to the public, then the document was redacted again and the yield numbers removed.

28 TNA DEFE 13/1478, British Nuclear Test Programme, V.H.B. Macklen, 19 April 1978.

29 TNA DEFE 13/1477, British Nuclear Test Programme, V.H.B. Macklen, 21 November 1977.

30 TNA DEFE 13/1478, British Nuclear Test Programme, 27 April 1978.

31 TNA DEFE 13-1478, British Nuclear Test Programme, V.H.B. Macklen, 19 April 1978. TNA DEFE 24/1361, Background Note, M. Nutting, 21 November 1978.

32 TNA DEFE 25/335, British Nuclear Test, V.H.B. Macklen, 20 November 1978.

33 TNA DEFE 19/181, Drake-Seager, ACO (W), British Embassy to V.H.B. Macklen, DCA (PN), MOD, Quargel – Schedule and Costs, 25 July 1978.

34 TNA DEFE 25/335, Strategic nuclear deterrent: Polaris Improvement Programme (PIP); Options, Phase 4, 6 June 1979, 21 August 1978.

35 TNA DEFE 24/1361, Background Note, M. Nutting, 21 November 1978.

36 TNA DEFE 24/1361, D.C. Fakley, D/DSc 6 to Head of DS2, 1980 Underground Nuclear Test Programme, 1 August 1979.

37 TNA DEFE 24/1361, AWRE Classification Notice No.43 Codenames of Underground Tests of UK Nuclear Devices at NTS, 7 March 1980.

38 TNA DEFE 24/1361, Dr A.M. Fox, Head of DS 2 to AUS (R), The Nuclear Test Programme, 14 September 1978.

39 TNA DEFE 24/1361, V.H.B. Macklen, DCA (PN) to Secretary of State, British Nuclear Test Programme, 2 October 1978.

40 TNA DEFE 24/1361, Secretary of State for Defence to Prime Minister, British Nuclear Test Programme, 3 November 1978.

41 TNA DEFE 19/181, Drake-Seager, Washington to Macklen, Nessel Meeting, 20 July 1978.

42 TNA DEFE 13/1477, The UK test on 16 March 1978, 30 November 1977.

43 TNA DEFE 13/1478, British Nuclear Test Programme, 3 November 1978.

44 TNA DEFE 13/1478, British Nuclear Test Programme, 3 November 1978.

45 TNA DEFE 13/1478, British Nuclear Test Programme, V.H.B. Macklen, 25 October 1978.

46 TNA DEFE 25/335, British Nuclear Test, V.H.B. Macklen, 20 November 1978.

47 TNA DEFE 19/181, Letter from V.H.B. Macklen, DCA (PN), 27 June 1979.

48 TNA PREM 19/14, R.L.L. Facer, MOD to B.G. Cartledge, 10 Downing Street, Comprehensive Test Ban (CTB), 11 May 1979.

49 TNA PREM 19/14, R.L.L. Facer, MOD to B.G. Cartledge, 10 Downing Street, Nuclear Warhead Test Programme, 24 May 1979.

50 TNA PREM 19/14, John Hunt, Cabinet Secretary to Prime Minister, 10 July 1979.

51 TNA PREM 19/14, Manuscript Note to Prime Minister, 29 August 1979.

52 TNA DEFE 24/1361, CSA to Secretary of State, British Underground Nuclear Test, 30 August 1979.

53 TNA DEFE 19/181, V.H.B. Macklen to PUS, British Nuclear Test Programme Note for the Record, 15 August 1978.

54 TNA DEFE 24/1361, AWRE Aldermaston Classification Notice No 40, Classification Guide for DICEL/NESSEL, 14 March 1979.
55 TNA DEFE 24/1361, V.H.B. Macklen, DCA (PN) to Secretary of State, British Nuclear Test Programme (undated autumn 1978). See also Defence Secretary to Prime Minister, British Nuclear Test Programme, 26 October 1978.
56 TNA DEFE 24/1361, British Nuclear Test Programme undated.
57 TNA DEFE 19/181, V.H.B. Macklen DCA (PN) to PUS, British Nuclear Test Programme, Note for the Record, 15 August 1978.
58 TNA PREM 19/14, R.L.L. Facer, MOD to B.G. Cartledge, 10 Downing Street, Nuclear Warhead Test Programme, 24 May 1979.
59 TNA PREM 19/159, Francis Pym to Prime Minister, British Underground Nuclear Test, 28 April 1980.
60 TNA PREM 19/14, R.L.L. Facer, MOD to B.G. Cartledge, 10 Downing Street, Nuclear Warhead Test Programme, 24 May 1979.
61 TNA PREM 19/159, Robert Armstrong to Prime Minister, Cabinet: British Nuclear Test Programme, 23 April 1980.
62 TNA DEFE 24/1361, AWRE Aldermaston Classification Notice no. 42 Classification Guide for DINGBAT/COLWICK, 12 December 1979.
63 TNA DEFE 24/1361, Defence Secretary to Prime Minister, British Nuclear Test Programme, 6 July 1979.
64 TNA CAB 128/68/13, Cabinet, CC (80), Conclusions of a Meeting of the Cabinet, p. 9, 23 October 1980.
65 TNA DEFE 19/181, BDS Washington to MODUK, 14 August 1979.
66 TNA CAB 164/1564, C.H.O'D. Alexander, Cabinet Office to D.C. Fakley, ACSA (N), MOD, Nuclear Test Ban Policy, 27 January 1981.
67 TNA PREM 19/159, N.J. Vile to Prime Minister, British Nuclear Test Programme, 9 January 1980.
68 TNA PREM 19/417, PPS 10 Downing Street to B.M. Norbury, MOD, British Nuclear Test Programme, 12 February 1981.
69 TNA PREM 19/417, J.D.S. Dawson, MOD to M.O.D.B. Alexander, 10 Downing Street, British Nuclear Test Programme, 17 December 1980.
70 TNA PREM 19/694, John Nott, Secretary of State for Defence to Prime Minister, British Underground Nuclear Test, 11 December 1981.
71 TNA PREM 19/694, Clive Whitmore, 10 Downing Street to David Omand, MOD, British Nuclear Test, 17 December 1981.
72 TNA PREM 19/417, Robert Armstrong to Prime Minister, British Nuclear Tests, 9 September 1980.
73 TNA DEFE 25/435, CSA to Secretary of State, British Underground Nuclear Test Programme, 7 October 1981.
74 TNA PREM 19/417, Lord Privy Seal to Prime Minister, British Underground Nuclear Test Programme, 9 September 1980.
75 TNA DEFE 25/435, CSA to Secretary of State, British Underground Nuclear Test Programme, 7 October 1981.
76 TNA PREM 19/417, Robert Armstrong to Prime Minister, British Nuclear Tests, 9 September 1980.
77 TNA PREM 19/695, D.T. Piper, MOD to A.J. Coles, 10 Downing Street, 29 April 1982.
78 TNA PREM 19/417, Carrington to Secretary of State for Defence, 15 June 1981.
79 TNA PREM 19/417, Robert Armstrong to Prime Minister, British Nuclear Tests, 9 September 1980.
80 TNA PREM 19/417, Clive Whitmore, 10 Downing Street to Brian Norbury, MOD, British Nuclear Test Programme, 10 June 1981.
81 TNA DEFE 13/2826/1, Annex A Proposal for an Underground UK Nuclear Tests in 1984, under draft minute from Dr J. A. Davies (Dr) Sc (Nuc)/1 to PS/CSA, British Underground Nuclear Test-1994, May 1992. There were three Trident validation tests, the last being in 1986 as Holbrook III.

Bibliography

Primary sources: United Kingdom
The National Archives, Kew.

AIR 2/19184 Future Nuclear Weapon System Requirements, 1970–1976.

AIR 2/20501 Progress of New Weapons and Future of an Airborne Nuclear Deterrent, 1972 Mar 10–1974 May 14.

AIR 8/2785 Stockpiling Policy and Planning: Includes Safety and Reliability of United Kingdom Nuclear Warheads Under a Comprehensive Test Ban, Assumptions for Logistic Planning, Review of War Material Reserves, 1972–1980.

AIR 8/3106 Nuclear Weapons, 1973 Sept 4–1977 Oct 14.

AIR 8/3116 Nuclear Weapons, 1973 Sept 11–1982 Feb 2.

AIR 8/3179 Suspension of Nuclear Tests, 1978 Oct 3–1982 Jan 21.

AIR 8/3362 Nuclear Weapons, 1982 Apr 18–1983 Sept 22.

AIR 20/12890 Vice-chief of Air Staff: Nuclear Weapons: Carriage and Storage, 1974 Jan 01–1975 Dec 31.

AVIA 65/2217 Agreements between the Minister of Aviation Supply and British Nuclear Fuels Ltd, 1970–1976.

AVIA 65/2436 Polaris Improvement Programme Chevaline (KH 793): UK/US Panton/ Barse Exchanges, 1972 Nov 23–1974 Aug 2.

CAB 130/720 International Aspects of Nuclear Defence Policy: Meetings 1–2, Papers 1–2, 1974 Mar 5–1974 May 29.

CAB 130/952 Official Group on International Aspects of Nuclear Defence Policy: Meetings 1–28, Papers 1–19, 1977 Feb 1–1977 Dec 21.

CAB 130/1011 Official Group on International Aspects of Nuclear Defence Policy: Meetings 1–26, papers 1–18, 1978 Jan 5–1978 Dec 5.

CAB 130/1072 Official Group on International Aspects of Nuclear Defence: Meetings 1–4, papers 1–5, 1979 Jan 11–1979 Mar 15.

CAB 130/1109 Nuclear Defence Policy: Meetings 1–4, 1979 May 24–1979 Dec 5.

CAB 130/1129 Nuclear Defence Policy: Meeting 1, papers 1–2, 1980.

CAB 130/1158 Official Group on International Aspects of Nuclear Defence Policy: Meeting 1, papers 1–3, 1981 Jan 12–1981 Feb 6.

CAB 130/1160 Nuclear Defence Policy: Meeting 1, paper 1, 1981.

CAB 130/1182 Nuclear Defence Policy: Meetings 1–4, papers 1–4, 1982.

CAB 134/3665 Official Committee on Nuclear Policy: General Sub-Committee: Papers 1–11 (1973), papers 1–4 (1974), 1973 Jan 15–1974 Feb 01.

CAB 134/3822 Official Committee on Nuclear Policy: Paper 1, 1974.

CAB 134/3823 Official Committee on Nuclear Policy: General Sub-Committee: Meetings 1–5, papers 1–17, 1974.

CAB 134/3950 Official Committee on Nuclear Policy: General Sub-Committee: Meetings 1–8, papers 1–25 (1975); papers 1–2 (1976), 1975 Jan 16–1976 Jan 27.

CAB 134/4052 Nuclear Policy (Official) Committee: Meetings 1–7, papers 1–39, 1976.

CAB 134/4066 Nuclear Policy Committee: Paper 1, 1976.

CAB 134/4067 Official Committee on Nuclear Policy: Paper 1, 1976.

CAB 134/4158 Nuclear Policy (Official) Committee: Meetings 1–6, papers 1–34, 1977.

CAB 134/4247 Nuclear Policy Official Committee: Meetings 1–9, papers 1–20, 1978 Jan 04–1978 Nov 23.

CAB 134/4248 Nuclear Policy Official Committee: Papers 21–37, 1978 May 12–1978 Nov 28.

CAB 134/4384 Nuclear Policy (Official) Committee: Meeting 1, papers 1–5 (1979); Meetings 1–3, papers 1–11 (1980), 1979 Jun 14–1980 Dec 05.

CAB 134/4386 Nuclear Policy (Official) Committee: Meetings 1–2, papers 1–12, 1979.

CAB 134/4559 Nuclear Policy (Official) Committee: Meetings 1–2, papers 1–4, 1981.

CAB 134/4640 Nuclear Policy (Official) Committee: Papers 1–2, 1982 Sep 15–1982 Oct 19.

CAB 164/1564 International Aspects of Nuclear Defence Policy: Comprehensive Test Ban (CTB), 1980 Jun 06–1981 Feb 05.

CAB 164/1664 International Aspects of Nuclear Defence Policy: Nuclear Arms Control, 1981 Feb 10–1983 May 13.

DEFE 11/791 General Nuclear Matters: Includes Soviet Suggestion for Comprehensive Test Ban Treaty, 1974 Feb 19–1977 Feb 21.

DEFE 11/793 Tactical Nuclear Weapons, 1972 Sep 05–1977 Nov 14.

DEFE 13/748 Resumption of Nuclear Tests, 1970–1974.

DEFE 13/1039 Nuclear Weapons Policy: POLARIS Improvement Policy, 1975 Sep 16–1976 Jun 25.

DEFE 13/1097 Nuclear Weapons, 1973 Jan 01–1974 Dec 31.

DEFE 13/1247 Nuclear Policy: Includes Safety Aspects and Policy on Testing, 1974 Jan 01–1975 Dec 31.

DEFE 13/1399 Nuclear Weapons Policy: Polaris Improvement and Future of the Strategic Nuclear Deterrent, 1973 Oct 01–1974 Mar 31.

DEFE 13/1768 Nuclear Weapons Policy: Polaris Improvements, 1976 Jul 6–1977 Jul 14.

DEFE 13/1769 Nuclear Weapons Policy: Polaris Improvements, 1977 Jul 13–1977 Oct 26.

DEFE 13/1871 Polaris and Trident: Improvements, 1982 Nov 10–1983 Feb 21.

DEFE 13/1945 Nuclear Weapon Policy: Polaris Improvement; Future of the Strategic National Deterrent, 1982 Jan 7–1982 Feb 15.

DEFE 13/1946 Nuclear Weapon Policy: Polaris Improvement (the Trident Decision), 1982 Mar 8–1982 Mar 16.

DEFE 13/2413 Comprehensive Test Ban, 1979 Jan 26–1979 Apr 25.

DEFE 13/2469 Nuclear Weapons-Test Ban Treaty and the Neutron Bomb, 1977 Jul 08–1977 Oct 31.

DEFE 13/2504 Nuclear Weapons: Policy; Polaris Improvements, Future of the Strategic Nuclear Deterrent (SND), 1981 Nov 25–1982 Jan 19.

DEFE 13/2826 National Nuclear Policy Nuclear Test Programme: Nuclear Testing; British Underground Nuclear Testing Programme; UK Nuclear Test Programme; Suspension of French Nuclear Tests; UK Involvement in US Underground Nuclear Test; Maralinga, 1988 Jun 17–1992 Jul 17.

DEFE 19/170 Chevaline: Polaris Re-entry Body Improvement Programme, 1974 Mar 19–1975 Sept 02.

DEFE 19/181 UK Underground Nuclear Tests, 1978 Jun 06–1979 Dec 11.

DEFE 19/181/1 Previously Closed Extracts: 14 Pages. Now Released and Reunited with Parent Piece.

DEFE 19/208 Ad Hoc Committees on Polaris Improvements, 1974 Mar 22–1975 Jan 09.

DEFE 19/230 Comprehensive Test Ban Treaty Negotiations: Verification Arrangements, 1979 Jan 01–1979 Dec 31.

DEFE 19/240 Comprehensive Test Ban Treaty Negotiations, 1978.

DEFE 19/241 Comprehensive Test Ban Treaty Negotiations, 1978.

DEFE 19/242 Comprehensive Test Ban Treaty Negotiations, 1978.

TNA DEFE 19/275 Duff-Mason Report, 1978 Dec.

DEFE 19/345 Presentation for Nuclear Matters Steering Group (MISC 7) on Trident, 1982 Feb 1–1982 Feb 12.

DEFE 23/218 Comprehensive Test Ban Treaty, 1978.

DEFE 23/288 Successor to Polaris: Trident D5 Decision, 1981 Dec 4–1982 Jan 6.

DEFE 24/1344 Comprehensive Test Ban Treaty, 1977.

DEFE 24/1361 Nuclear Tests Approvals, 1978 Jan 01–1980 Dec 31.

DEFE 24/1819 Storage, Transportation and Safety Arrangements for Nuclear Weapons, 1976 Jan 01–1977 Dec 31.

DEFE 24/2116 Strategic Nuclear Deterrent: Successor System for Polaris and Chevaline, 1980 Jan 01–1980 Dec 31.

DEFE 24/2125 Nuclear Policy: UK Strategic Nuclear Deterrent (Trident C-4 Decision), 1980 Jan 01–1980 Dec 31.

DEFE 24/3010 Nuclear Weapons Deployment and Security: Deployment of UK Weapons, 1981 Nov 01–1983 Nov 30.

DEFE 25/595 Nuclear Weapons, 1981 Feb 01–1982 Oct 31.

DEFE 68/918 Arms Limitation: Comprehensive Test Ban (CTB); Nuclear Test Ban, 1982 Feb 9–1985 May 10.

DEFE 69/464 Defence Review 1974: Nuclear Weapons, 1974 Jan 01–1974 Dec 31.

DEFE 69/741 Chevaline: Improved Front End (IFE); Introduction of First Rate (Lighter Warheads), 1977 Aug 01–1978 Oct 31.

DEFE 69/742 Chevaline: Improved Front End (IFE); Introduction, 1978 Oct 01–1978 Oct 31.

DEFE 69/1265 Polaris: Chevaline, 1977 Jul 01–1982 Aug 31.

DEFE 69/1289 Trident in Service Date: Implication of Delay, 1981 Apr 01–1982 Mar 31.

DEFE 69/1307 Nuclear Weapons Policy: Theatre Nuclear Weapons Policy Steering Group, 1982 Jan 01–1983 Aug 31.

DEFE 69/1312 Chevaline Missile System: Approval for Service and Deployment, 1981 Sep 01–1982 Aug 31.

DEFE 69/1655 A3TK (Chevaline): Acceptance Into Service, 1976 Mar 1–1978 Feb 14.

DEFE 70/433 Disarmament: Cruise Missiles, Strategic Arms Limitation Talks (SALT), Comprehensive Test Ban Treaty, 1976 Nov 01–1977 Jul 31.

DEFE 72/167 Move of Facility for Manufacture of Lithium Salts for Use in Nuclear Weapons Programme from ROF Chorley to ROF Burghfield, 1977 Oct 01–1979 Nov 30.

DEFE 72/302 Theatre Nuclear Weapons Policy Steering Group, 1980 Jun 24–1982 Oct 8.

DEFE 72/512 Chevaline: Acceptance into Service, 1975 Jul 1–1976 Dec 21.

DEFE 72/523 Trident: Policy Group, 1980 Aug 1–1980 Aug 10.

DEFE 72/545 Chevaline: Project Review Committee, 1977 May 10–1978 May 12.

DEFE 72/569 Trident: General, 1980 8 Jul–1980 Sept 10.

DEFE 72/616 Chevaline: Production Aspects, 1975 Nov 10–1981 Sept 28.

ES 12/675 Annual Review in Support of the CHEVALINE Programme, 1980.

ES 14/62 Desirable Characteristics of Tactical Nuclear Weapons Systems, 1974 Jan 16–1974 Jan 16.

ES 15/406 Defence Procurement (Nuclear) (DPN) Board of Management: Agendas and Minutes, 1979 Jan 01–1979 Dec 31.

ES 15/407 Defence Procurement (Nuclear) (DPN) Board of Management: Agendas and Minutes, 1980 Jan 01–1980 Dec 31.

ES 16/33 Visit of L Turnbull and W Heckrotte to AWRE (Blacknest), 8 November 1977: Seismological Verification of Comprehensive Test Ban (CTB), 1977 Jan 01–1977 Dec 31.

ES 16/100 Photographs of the Banon Event at the Nevada Test Site, 1976 Jan 01–1976 Dec 31.

ES 16/102 Photographs of the FONDUTTA Event at the Nevada Test Site, 1978 Jan 01–1978 Dec 31.

ES 22/6 General Background Information on Nuclear Weapons and Tests, 1980 Jan 01–2008 Dec 31.

ES 24 Atomic Weapons Research Establishment and Atomic Weapons Establishment: Monitoring of International Tests: Files and Reports, 1958–1976.

ES 24/5 Comprehensive Test Ban Treaty: Policy, 1958 Jan 1–1976 Dec 31.

ES 25/3 Underground Sites for UK Nuclear Tests, 1972.

FCO 31/2458 Comprehensive Test Ban (CTB) Negotiations: Proposed National Seismic Stations (NSS) in UK Dependent Territories, 1978.

FCO 40/1044 Comprehensive Test Ban Treaty: National Seismic Stations on UK Dependent Territories, 1979.

FCO 46/1566 Defence Aspects of Nuclear Tests, 1977.

FCO 46/1567 Defence Aspects of Nuclear Tests, 1977.

FCO 46/1829 Defence Aspects of Nuclear Tests, 1978.

FCO 46/1830 Defence Aspects of Nuclear Tests, 1978.

FCO 46/2286 Comprehensive Test Ban (CTB), 1980.

FCO 46/2287 UK Strategic Deterrent: Decision to Replace Polaris with Trident, 1980.

FCO 46/2288 UK Strategic Deterrent: Decision to Replace Polaris with Trident, 1980.

FCO 46/2289 UK Strategic Deterrent: Decision to Replace Polaris with Trident, 1980.

FCO 46/2290 UK Strategic Deterrent: Decision to Replace Polaris with Trident, 1980.

FCO 46/2750 The Trident Programme, 1981.

FCO 46/2751 The Trident Programme, 1981.

FCO 46/2753 The Trident Programme, 1981.

FCO 46/2758 Comprehensive Test Ban (CTB), 1978.

FCO 46/3103 The Trident Programme, 1982.

FCO 46/3104 The Trident Programme, 1982.

FCO 46/3105 The Trident Programme, 1982.

FCO 46/3108 The Trident Programme, 1982.

FCO 46/3109 The Trident Programme, 1982.

FCO 46/3110 The Trident Programme, 1982.

FCO 66/615 Negotiations for a Comprehensive Nuclear Test Ban Treaty, 1974.

FCO 66/616 Negotiations for a Comprehensive Nuclear Test Ban Treaty, 1974.

FCO 66/617 Negotiations for a Comprehensive Test Ban Treaty, 1974.

FCO 66/618 Negotiations for a Comprehensive Test Ban Treaty, 1974.

FCO 66/675 Peaceful Nuclear Explosions (PNEs): International Control, 1974.

FCO 66/676 Peaceful Nuclear Explosions (PNEs): International Control, 1974.

FCO 66/677 Peaceful Nuclear Explosions (PNEs): International Control, 1974.

FCO 66/678 Peaceful Nuclear Explosions (PNEs): International Control, 1974.

FCO 66/679 Peaceful Nuclear Explosions (PNEs): International Control, 1974.

FCO 66/680 Peaceful Nuclear Explosions (PNEs): International Control, 1974.

FCO 66/773 Negotiations for a Comprehensive Nuclear Test Ban, 1975.

FCO 66/774 Negotiations for a Comprehensive Nuclear Test Ban, 1975.

FCO 66/775 Negotiations for a Comprehensive Nuclear Test Ban, 1975.

FCO 66/806 Discussions on Peaceful Nuclear Explosions (PNEs) in Conference of Committee on Disarmament (CCD), 1975.

FCO 66/807 Discussions on Peaceful Nuclear Explosions (PNEs) in Conference of Committee on Disarmament (CCD), 1975.

FCO 66/808 Peaceful Nuclear Explosions Committee in International Atomic Energy Agency (IAEA), 1975.

FCO 66/809 Peaceful Nuclear Explosions Committee in International Atomic Energy Agency (IAEA), 1975.

FCO 66/810 Peaceful Nuclear Explosions Committee in International Atomic Energy Agency (IAEA), 1975.

FCO 66/811 Peaceful Nuclear Explosions Committee in International Atomic Energy Agency (IAEA), 1975.

FCO 66/812 Peaceful Nuclear Explosions Committee in International Atomic Energy Agency (IAEA), 1975.

FCO 66/813 Peaceful Nuclear Explosions Committee in International Atomic Energy Agency (IAEA), 1975.

FCO 66/814 Peaceful Nuclear Explosions (PNEs): International Control; International Atomic Energy Association (IAEA) Technical Work, 1975.

FCO 66/815 Peaceful Nuclear Explosions (PNEs): International Control; International Atomic Energy Association (IAEA) Technical Work, 1975.

FCO 66/816 Peaceful Nuclear Explosions (PNEs): International Control; International Atomic Energy Association (IAEA) Technical Work, 1975.

FCO 66/874 Negotiations for a Comprehensive Test Ban (CTB): Bilateral Discussions between USA and Soviet Union, 1977.

FCO 66/875 Negotiations for a Comprehensive Test Ban (CTB): Tripartite Discussions between UK, USA and Soviet Union, Jul 1977; briefs 1977.

FCO 66/876 Negotiations for a Comprehensive Test Ban (CTB): Tripartite Discussions between UK, USA and Soviet Union, Jul 1977; briefs 1977.

FCO 66/877 Negotiations for a Comprehensive Test Ban (CTB): Views of Fourth Countries, 1977.

FCO 66/878 Negotiations for a Comprehensive Test Ban (CTB): Bilateral Discussions between UK and USA, 1977.

FCO 66/879 Negotiations for a Comprehensive Test Ban (CTB): Verification Arrangements, 1977.

FCO 66/880 Negotiations for a Comprehensive Test Ban (CTB): Bilateral Discussions between UK and Soviet Union, 1977.

FCO 66/881 Negotiations for a Comprehensive Test Ban Treaty (CTBT): Tripartite Discussions between UK, USA and Soviet Union, Phase 1, 1977 Jul.

FCO 66/882 Negotiations for a Comprehensive Test Ban (CTB): Briefing of Fourth Countries and International Organisations, 1977.

FCO 66/883 Negotiations for a Comprehensive Test Ban (CTB): Peaceful Nuclear Explosions (PNE), 1977.

FCO 66/884 Negotiations for a Comprehensive Test Ban (CTB): Briefing of UK Ministers and Officials, 1977.

FCO 66/885 Contacts between UK and USA Following Phase 1 of Comprehensive Test Ban (CTB) Negotiations, 1977.

FCO 66/886 Negotiations for a Comprehensive Test Ban (CTB): Tripartite Discussions between UK, USA and Soviet Union, Phase 2, 1977 Oct.

FCO 66/887 Negotiations for a Comprehensive Test Ban (CTB): Second Round of Bilateral Discussions between UK and USA, 1977.

FCO 66/888 Comprehensive Test Ban (CTB): Verification; On-site Inspections, 1977.

FCO 66/889 Comprehensive Test Ban (CTB): Ministerial Exchanges between UK and Soviet Union, 1977.

FCO 66/890 Comprehensive Test Ban (CTB): Entry into Force, 1977.

FCO 66/891 Comprehensive Test Ban (CTB) and the United Nations General Assembly, 1977.

FCO 66/892 Comprehensive Test Ban (CTB): Tripartite Meetings of Political Advisers, 1977.

FCO 66/893 Negotiations for a Comprehensive Test Ban (CTB): Second Round Conversations with Soviet Officials, 1977.

FCO 66/894 Negotiations for a Comprehensive Test Ban (CTB): Tripartite Meetings on Technical Matters, 1977.

FCO 66/895 Negotiations for a Comprehensive Test Ban (CTB): Policy of Soviet Union, 1977.

FCO 66/896 Negotiations for a Comprehensive Test Ban (CTB): Policy of USA, 1977.

FCO 66/897 UK Policy on a Comprehensive Test Ban (CTB), 1977.

FCO 66/898 UK Policy on a Comprehensive Test Ban (CTB), 1977.

FCO 66/899 UK Policy on a Comprehensive Test Ban (CTB), 1977.

FCO 66/900 UK Policy on a Comprehensive Test Ban (CTB), 1977.

FCO 66/901 UK Policy on a Comprehensive Test Ban (CTB), 1977.

FCO 66/902 UK Policy on a Comprehensive Test Ban (CTB), 1977.

FCO 66/903 Negotiations for a Comprehensive Test Ban (CTB): Publicity, 1977.

FCO 66/921 Comprehensive Test Ban (CTB): Four Power Talks between UK, USA, Soviet Union and France, 1977.

FCO 66/1022 Comprehensive Test Ban (CTB): Tripartite Discussions between UK, USA and Soviet Union, 1978.

FCO 66/1023 Comprehensive Test Ban (CTB): Tripartite Discussions between UK, USA and Soviet Union, 1978.

FCO 66/1024 Comprehensive Test Ban (CTB): Tripartite Discussions between UK, USA and Soviet Union, 1978.

FCO 66/1025 Comprehensive Test Ban (CTB): Bilateral Discussions between UK and USA, 1978.

FCO 66/1026 Comprehensive Test Ban (CTB): Bilateral Discussions between UK and USA, 1978.

FCO 66/1027 Comprehensive Test Ban (CTB): Bilateral Discussions between UK and USA, 1978.

FCO 66/1028 Comprehensive Test Ban (CTB): Peaceful Nuclear Explosions, 1978.

FCO 66/1029 Comprehensive Test Ban (CTB): Verification; On-site Inspection (OSI), 1978.

FCO 66/1030 Comprehensive Test Ban (CTB): Verification: Internal Seismic Stations (ISS), Regional Seismic Installations (RSI) and National Seismic Stations (NSS), 1978.

FCO 66/1031 Comprehensive Test Ban (CTB): Verification: Internal Seismic Stations (ISS), Regional Seismic Installations (RSI) and National Seismic Stations (NSS), 1978.

FCO 66/1032 Comprehensive Test Ban (CTB): Form of Treaty, 1978.

FCO 66/1033 Comprehensive Test Ban (CTB) Treaty: Form of the Ultimate Tripartite Documents, 1978.

FCO 66/1034 Comprehensive Test Ban (CTB): Bilateral Discussions between USA and Soviet Union, 1978.

FCO 66/1035 Comprehensive Test Ban (CTB): Conversations between UK and Soviet Union, 1978.

FCO 66/1036 Comprehensive Test Ban (CTB): Views of Fourth Countries, 1978.

FCO 66/1037 Comprehensive Test Ban (CTB): Briefing of Fourth Countries and International Organisations, 1978.

FCO 66/1038 Comprehensive Test Ban (CTB): Round-ups and State of Play Reports, 1978.

FCO 66/1039 Comprehensive Test Ban (CTB): The Two Track Approach and UK Participation in the 'Inner Track', 1978.

FCO 66/1040 Comprehensive Test Ban (CTB): The Two Track Approach and UK Participation in the 'Inner Track', 1978.

FCO 66/1041 Comprehensive Test Ban (CTB) Technical Bilateral Discussions between UK and USA, 1978 (Plus FCO 66/1041/1).

FCO 66/1042 Comprehensive Test Ban (CTB): Duration of Treaty, 1978.

FCO 66/1043 Comprehensive Test Ban (CTB) and United Nations General Assembly (UNGA), 1978.

FCO 66/1044 Comprehensive Test Ban (CTB): Soviet Press Comment, 1978.

FCO 66/1045 Comprehensive Test Ban (CTB): Definitions; United States Position, 1978.

FCO 66/1046 Comprehensive Test Ban (CTB): Definitions; United States Position, 1978.

FCO 66/1047 Comprehensive Test Ban (CTB): Definitions; United States Position, 1978.

FCO 66/1048 Comprehensive Test Ban (CTB): Definitions; United States Position, 1978.

FCO 66/1049 Comprehensive Test Ban (CTB): Definitions; United States Position, 1978.

FCO 66/1050 Comprehensive Test Ban (CTB) and the Conference of the Committee on Disarmament (CCD), 1978.

FCO 66/1051 Comprehensive Test Ban (CTB) and the Conference of the Committee on Disarmament (CCD), 1978.

FCO 66/1052 Comprehensive Test Ban (CTB): International Consultative Commission (ICC), 1978.

FCO 66/1053 Comprehensive Test Ban (CTB): International Exchange of Seismic Data, 1978.

FCO 66/1054 Comprehensive Test Ban (CTB): International Atomic Energy Agency (IAEA) Work on Peaceful Nuclear Explosions, 1978.

FCO 66/1055 Comprehensive Test Ban (CTB): Records of the Tripartite Working Group on Technical Matters, 1978.

FCO 66/1056 Comprehensive Test Ban (CTB): Records of the Tripartite Working Group on Technical Matters, 1978.

FCO 66/1057 Comprehensive Test Ban (CTB): Records of the Tripartite Working Group on Technical Matters, 1978.

FCO 66/1058 Comprehensive Test Ban (CTB): Briefing of UK Ministers and Officials, 1978.

FCO 66/1059 Comprehensive Test Ban (CTB): Records of the Tripartite Political Working Group, 1978.

FCO 66/1060 Comprehensive Test Ban (CTB): Verification, 1978.

FCO 66/1061 Comprehensive Test Ban (CTB): Personalities; Members of Delegations, 1978.

FCO 66/1062 Comprehensive Test Ban (CTB): Records of Senior Political Advisers' Meetings, 1978.

FCO 66/1063 Relationship of a Comprehensive Test Ban (CTB) Treaty to Other Treaties, 1978.

FCO 66/1064 Comprehensive Test Ban (CTB): The Ambiguous Event Problem (the Seismic Event Problem), 1978.

FCO 66/1065 Comprehensive Test Ban (CTB): The Separate Agreement; the Two Track Approach, 1978.

FCO 66/1066 Messages between James Callaghan, UK Prime Minister, and Jimmy Carter, US President, on Comprehensive Test Ban (CTB) Negotiations, 1978.

FCO 66/1067 Moratorium on Nuclear Explosions Pending Conclusion of a Comprehensive Test Ban (CTB), 1978.

FCO 66/1068 Comprehensive Test Ban (CTB): Treaty Language; Final Clauses Excluding Duration, 1978.

FCO 66/1069 Comprehensive Test Ban (CTB) and Weapons Testing, 1978.

FCO 66/1070 Comprehensive Test Ban (CTB): Chemical Explosions, 1978.

FCO 66/1071 Comprehensive Test Ban (CTB): Language of Article 1, Basic Prohibition, 1978.

FCO 66/1072 Comprehensive Test Ban (CTB): Permitted Experiments, 1978.

FCO 66/1073 Comprehensive Test Ban (CTB): Permitted Experiments, 1978.

FCO 66/1074 Comprehensive Test Ban (CTB): The Post Trilateral Phase; Procedural Aspects, Implications for Adherence of Non-nuclear Weapon States (NNWS), 1978.

FCO 66/1075 Comprehensive Test Ban (CTB): Treaty Language; Review Conference, 1978.

FCO 66/1076 Comprehensive Test Ban (CTB): Independent Panel of Scientists, 1978.

FCO 66/1077 Comprehensive Test Ban (CTB): Policy of Soviet Union, 1978.

FCO 66/1078 Comprehensive Test Ban (CTB): Policy of USA, 1978.

FCO 66/1079 Attitude of United States Congress to Comprehensive Test Ban (CTB) Treaty, 1978.

FCO 66/1080 Comprehensive Test Ban (CTB): UK Policy, 1978.

FCO 66/1081 Polaris Nuclear Testing Improvement Programme in Connection with a Comprehensive Test Ban (CTB), 1978.

FCO 66/1082 Polaris Nuclear Testing Improvement Programme in Connection with a Comprehensive Test Ban (CTB), 1978.

FCO 66/1083 Comprehensive Test Ban (CTB): Publicity, 1978.

FCO 66/1084 Comprehensive Test Ban (CTB): Publicity, 1978.

FCO 66/1116 International Atomic Energy Agency (IAEA): Peaceful Nuclear Explosions, 1968.

FCO 66/1266 Comprehensive Test Ban (CTB): Tripartite Negotiations between the UK, USA and Soviet Union; Plenary and Restricted Meetings, 1979.

FCO 66/1267 Comprehensive Test Ban (CTB): Tripartite Negotiations between the UK, USA and Soviet Union; Meetings of Political Working Group, 1979.

FCO 66/1268 Comprehensive Test Ban (CTB): Tripartite Negotiations between the UK, USA and Soviet Union; Meetings of Technical Working Group and On-Site Inspection (OSI) Sub-Group, 1979.

FCO 66/1269 Comprehensive Test Ban (CTB): Bilateral Discussions between the Delegations of the UK and USA, 1979.

FCO 66/1270 Comprehensive Test Ban (CTB): Bilateral Discussions between the Delegations of the UK and Soviet Union, 1979.

FCO 66/1271 Comprehensive Test Ban (CTB): Bilateral Discussions between the Delegations of the USA and Soviet Union, 1979.

FCO 66/1272 Comprehensive Test Ban (CTB): Non-delegation Bilateral Discussions between the UK and USA, 1979.

FCO 66/1273 Comprehensive Test Ban (CTB): Non-delegation Bilateral Discussions between the UK and USA, 1979.

FCO 66/1274 Comprehensive Test Ban (CTB): Non-delegation Bilateral Discussions between the UK and USA, 1979.

FCO 66/1275 Comprehensive Test Ban (CTB): Non-delegation Bilateral Discussions between the UK and Soviet Union, 1979.

FCO 66/1276 Comprehensive Test Ban (CTB): Non-delegation Bilateral Discussions between the USA and Soviet Union, 1979.

FCO 66/1277 Comprehensive Test Ban (CTB) Negotiations: Round-ups and State of Play Reports, 1979.

FCO 66/1278 Comprehensive Test Ban (CTB): Views of Fourth Countries, 1979.

FCO 66/1279 Comprehensive Test Ban (CTB): Briefing of Fourth Countries and International Organisations, 1979.

FCO 66/1280 Comprehensive Test Ban (CTB): National Seismic Stations, 1979.

FCO 66/1281 Comprehensive Test Ban (CTB): National Seismic Stations, 1979.

FCO 66/1282 Comprehensive Test Ban (CTB): National Seismic Stations, 1979.

FCO 66/1283 Comprehensive Test Ban (CTB): National Seismic Stations, 1979.

FCO 66/1284 Comprehensive Test Ban (CTB): Review Conference Options, 1979.

FCO 66/1285 Comprehensive Test Ban (CTB) and the Committee on Disarmament, 1979.

FCO 66/1286 Comprehensive Test Ban (CTB): Permitted Experiments, 1979.

FCO 66/1287 Comprehensive Test Ban (CTB): On-site Inspection; Rights, Privileges and Immunities, 1979.

FCO 66/1288 Comprehensive Test Ban (CTB): Seismic Experts Group of the Committee on Disarmament, 1979.

FCO 66/1289 Comprehensive Test Ban (CTB): Personalities; Members of Delegations, 1979.

FCO 66/1290 Comprehensive Test Ban (CTB): Briefs, 1979.

FCO 66/1291 Comprehensive Test Ban (CTB): Negotiating Timetable, 1979.

FCO 66/1292 Comprehensive Test Ban (CTB): Papers for Official Group in International Aspects of Nuclear Defence Policy (GEN 63 and MISC 1), 1979.

FCO 66/1293 Comprehensive Test Ban (CTB): Separate Verification Agreement (SVA): Joint Consultative Commission (JCC), 1979.

FCO 66/1294 Comprehensive Test Ban (CTB) and the United Nations General Assembly (UNGA), 1979.

FCO 66/1295 Comprehensive Test Ban (CTB): UK Role in the Separate Verification Agreement (SVA), 1979.

FCO 66/1296 Comprehensive Test Ban (CTB): Consultations with UK Dependent Territories on National Seismic Stations, 1979.

FCO 66/1297 Comprehensive Test Ban (CTB): Consultations with UK Dependent Territories on On-site Inspection, 1979.

FCO 66/1298 Comprehensive Test Ban (CTB): International Exchange of Seismic Data, 1979.

FCO 66/1299 Comprehensive Test Ban (CTB): Peaceful Nuclear Explosions, 1979.

FCO 66/1300 Comprehensive Test Ban (CTB) and Non-proliferation, 1979.

FCO 66/1301 Comprehensive Test Ban (CTB): Visit of UK and Soviet Union Specialists to National Seismic Stations in the USA, 1979 Aug.

FCO 66/1302 Comprehensive Test Ban (CTB): Visit of President Carter's Scientific Adviser, Dr Frank Press, to the UK, 1979 Jun.

FCO 66/1303 Comprehensive Test Ban (CTB): United States Testing Programme, 1979.

FCO 66/1304 Comprehensive Test Ban (CTB): Treaty Preamble, 1979.

FCO 66/1305 Comprehensive Test Ban (CTB): Consultations between the UK and USA on a Threshold Test Ban, 1979.

FCO 66/1306 Comprehensive Test Ban (CTB): Policy of the Soviet Union, 1979.

FCO 66/1307 Comprehensive Test Ban (CTB): Policy of the USA, 1979.

FCO 66/1308 Comprehensive Test Ban (CTB): Visit of Dr Herbert York, Leader of the US Delegation, to the UK, 1979 Feb.

FCO 66/1309 Comprehensive Test Ban (CTB): Attitude of the United States Congress, 1979.

FCO 66/1310 Comprehensive Test Ban (CTB): UK Policy Review; Possible New Approaches, 1979.

FCO 66/1311 Comprehensive Test Ban (CTB): UK Nuclear Tests, 1979.

FCO 66/1312 Comprehensive Test Ban (CTB) Negotiations: Resumptions and Recesses, 1979.

FCO 66/1313 Comprehensive Test Ban (CTB): Publicity, 1979.

FCO 66/1460 Comprehensive Test Ban (CTB): Tripartite Negotiations between the UK, USA and Soviet Union; Plenary and Restricted Meetings, 1980.

FCO 66/1461 Comprehensive Test Ban (CTB): Tripartite Negotiations between the UK, USA and Soviet Union; Meetings of Political Working Group and Political Advisers, 1980.

FCO 66/1462 Comprehensive Test Ban (CTB): Tripartite Negotiations between the UK, USA and Soviet Union; Meetings of Technical Working Group and On-Site Inspection (OSI) Sub-Group, 1980.

FCO 66/1463 Comprehensive Test Ban (CTB): Bilateral Discussions between the Delegations of the UK and USA, 1980.

FCO 66/1464 Comprehensive Test Ban (CTB): Bilateral Discussions between the Delegations of the UK and Soviet Union, 1980.

FCO 66/1465 Comprehensive Test Ban (CTB): Non-delegation Bilateral Discussions between the UK and USA, 1980.

FCO 66/1466 Comprehensive Test Ban (CTB): Non-delegation Bilateral Discussions between the UK and Soviet Union, 1980.

FCO 66/1467 Comprehensive Test Ban (CTB): National Seismic Stations, 1980.

FCO 66/1468 Comprehensive Test Ban (CTB): The Multilateral Dimension, 1980.

FCO 66/1469 Comprehensive Test Ban (CTB): Installation of Prototype National Seismic Stations, 1980.

FCO 66/1470 Comprehensive Test Ban (CTB): National Seismic Stations on UK Dependent Territories, 1980.

FCO 66/1471 Comprehensive Test Ban (CTB): The Negotiating Timetable, 1980.

FCO 66/1472 Comprehensive Test Ban (CTB): Tripartite Reports to the Committee on Disarmament, 1980.

FCO 66/1473 UK Policy on a Comprehensive Test Ban (CTB) Treaty, 1980.

FCO 66/1474 UK Policy on a Comprehensive Test Ban (CTB) Treaty, 1980.

FCO 66/1475 UK Policy on a Comprehensive Test Ban (CTB) Treaty, 1980.

FCO 66/1532 Comprehensive Test Ban (CTB): Policy Review, 1981.

FCO 66/1533 Comprehensive Test Ban (CTB): Policy Review, 1981.

FCO 66/1534 Comprehensive Test Ban (CTB): Bilateral Meetings between the UK and USA, 1981.

FCO 66/1539 Nuclear Tests, 1981.

FCO 66/1586 Comprehensive Test Ban (CTB): Policy, 1982.

FCO 66/1587 Comprehensive Test Ban (CTB): Policy, 1982.

FCO 66/1588 Comprehensive Test Ban (CTB): Policy, 1982.

FCO 66/1589 Comprehensive Test Ban (CTB): Policy, 1982.

FCO 66/1590 Comprehensive Test Ban (CTB): Policy, 1982.

FCO 66/1591 Comprehensive Test Ban (CTB): Verification Papers, 1982.

FCO 160/55/30 'Negotiations for a Comprehensive Ban on Nuclear Tests', Despatch from Percy Cradock, Head of the UK Delegation to Comprehensive Test Ban Discussions, Geneva, 1978.

FCO 160/206/29 'Nuclear Test Ban Negotiations 1977–1980: Part I – Where Are We?', Despatch from John Christopher Edmonds, HM Ambassador and Leader of the UK Delegation to the Comprehensive Test Ban Treaty Negotiations, Geneva, 1978.

FCO 160/206/30 'Nuclear Test Ban Negotiations 1977–1980: Part II – What Now?', Despatch from John Christopher Edmonds, HM Ambassador and Leader of the UK Delegation to the Comprehensive Test Ban Treaty Negotiations, Geneva, 1980.

PREM 16/93 FOREIGN POLICY. Soviet/US Threshold Treaty on Underground Nuclear Weapons Tests, 1974 Jun 15–1974 Oct 31.

PREM 16/1183 DEFENCE. Mutual and Balanced Reduction of Forces (MBFR): Strategic Arms Limitation Talks (SALT); Comprehensive Test Ban Negotiations with USA and USSR in Geneva, Jul 1977, part 1, 1974 Jun 13–1977 Jun 29.

PREM 16/1184 DEFENCE. Mutual and Balanced Reduction of Forces (MBFR): Strategic Arms Limitation Talks (SALT); Comprehensive Test Ban Negotiations with USA and USSR in Geneva, Jul 1977, Part 2, 1977 Jul 07–1977 Nov 25.

PREM 16/1570 DEFENCE. Mutual and Balanced Reduction of Forces (MBFR): Strategic Arms Limitation Talks (SALT); Convention on Chemical Weapons; Comprehensive Test Ban Negotiations, Part 3, 1977 Nov 23–1978 Apr 21.

PREM 16/1571 DEFENCE. Mutual and Balanced Reduction of Forces (MBFR): Strategic Arms Limitation Talks (SALT); Convention on Chemical Weapons; Comprehensive Test Ban negotiations, Part 4, 1978 Apr 24–1978 Jul 10.

PREM 16/1572 DEFENCE. Mutual and Balanced Reduction of Forces: Strategic Arms Limitation Talks; Convention on Chemical Weapons; Comprehensive Test Ban Negotiations, Part 5, 1978 Jul 12–1978 Oct 23, 1978 Oct 24–1979 Mar 14.

PREM 16/1979 DEFENCE. Mutual and Balanced Force Reduction (MBFR) Discussions; SALT II and III Discussions; Convention on Chemical Weapons; Negotiations on Comprehensive Test Ban, Part 6, 1978 Oct 24–1979 Mar 14.

PREM 16/1980 DEFENCE. Mutual and Balanced Force Reduction (MBFR) Discussions; SALT II and III Discussions; Convention on Chemical Weapons; Negotiations on Comprehensive Test Ban, Part 7, 1979 Mar 21–1979 May 01.

PREM 19/212 DEFENCE. Comprehensive Test Ban (CTB) and Strategic Arms Limitation Talks (SALT); Mutual and Balanced Force Reductions (MBFR), Part 1, 1979 May 04–1979 Jun 15.

PREM 19/213 DEFENCE. Comprehensive Test Ban (CTB) and Strategic Arms Limitation Talks (SALT); Mutual and Balanced Force Reductions (MBFR), Part 2, 1979 Jun 18–1979 Aug 24.

PREM 19/693 DEFENCE. Mutual and Balanced Force Reductions (MBFR) Discussions; SALT II and III Discussions; Convention on Chemical Weapons; Negotiations on Comprehensive Test Ban, Part 3, 1979 Sep 3–1982 Dec 24.

PREM 19/695 DEFENCE. UK Strategic Nuclear Deterrent; Trident Missile Programme, Part 5, 1982 Mar 1–1982 Sep 09.

PREM 19/2930 DEFENCE. Arms Control and the Effect on Nuclear Arms: Strategic Arms Limitation Talks (SALT); Mutual and Balanced Force Reductions (MBFR) Discussions; Nuclear Advisory Panel; Negotiations on Comprehensive Test Ban, Part 17, 1979 Mar 2–1989 May 4.

UK CCD Working Papers

Working Paper on a Development in Discriminating between Seismic Sources, submitted by the United Kingdom on 13 Aug 1974, CCD/440.

Working Paper on Safeguards against the Employment of Multiple Explosion to Simulate Earthquakes, submitted by the United Kingdom on 24 July 1975, CCD/459.

Working Paper on the Processing and Communication of Seismic Data to Provide for National Means of Verifying a Test Ban, submitted by the United Kingdom on 12 Apr 1976, CCD/487 (and Corr.1).

Working Paper on the Recording and Processing of P Waves to Provide Seismograms Suitable for Discriminating between Earthquakes and Underground Explosions Submitted by the United Kingdom on 12 Apr 1976, CCD/488.

Working Paper on Safeguards against the Employment of Multiple Explosion to Simulate Earthquakes, submitted by the United Kingdom on 24 Jul 1975, CCD/459.

Working Paper on the United Kingdom's Contribution to Research on Seismological Problems Relating to Underground Nuclear Tests, submitted by the United Kingdom on 12 Apr 1976, CCD/486 (and Corr.1).

Primary Sources: the US

Foreign Relations of the United States, 1977–1980, Volume XXVI, Arms Control and Nonproliferation, Washington, DC, 2015.

National Security Archive, The George Washington University, Washington, DC.

Secondary sources

Arnold, Lorna and Mark Smith, *Britain, Australia and the Bomb The Nuclear Tests and Their Aftermath*, Basingstoke, Palgrave Macmillan, 2006.

Baylis, John and Kristan Stoddard, *The British Nuclear Experience: The Roles of Beliefs, Culture and Identify*, Oxford, Oxford University Press, 2014.

Dahlman, Ola, Svein Mykkeltveit, and Hein Haak, *Nuclear Test Ban: Converting Political Visions to Reality*, Dordrecht, Springer, 2009.

Dillon, G. M., *Dependence and Deterrence, Success and Civility in the Anglo-American Special Nuclear Relationship, 1962–1982*, Aldershot, Dartmouth Publishing Co. Ltd, 1983.

Douglas, Alan, *Forensic Seismology and Nuclear Test Bans*, Cambridge, Cambridge University Press, 2013.

Fetter, Steve, *Toward a Comprehensive Test Ban*, Cambridge, Ballinger, 1988.

Freedman, Lawrence, *Britain and Nuclear Weapons*, London, Macmillan, 1980.

Garthoff, Raymond, *Détente and Confrontation: American and Soviet Relations from Nixon to Reagan*, Washington, DC, Brookings Institution, 1985.

Goldblat, Jozef and David Cox, *Nuclear Weapon Tests: Prohibition or Limitation?* (SIPRI Monographs), Oxford, Oxford University Press, 1988.

Hall, Keith, *Polaris The History of the UK's Submarine Force*, Stroud, The History Press, 2018.

Hawkings, David, *Keeping the Peace: The Aldermaston Story*, London, Leo Cooper, AWE, 2000.

Hennessy, Peter and James Jinks, *The Silent Deep The Royal Navy Submarine Service since 1945*, London, Penguin Books, 2016.

Hoffman, Mark, editor, *United Kingdom Arms Control Policy in the 1990's*, Manchester, Manchester University Press, 1990.

Jones, Matthew, *The Official History of the UK Strategic Nuclear Deterrent, Volume I: From the V-Bomber Era to the Arrival of Polaris, 1954–1964*, Abingdon, Routledge, 2017.

Jones, Matthew, *The Official History of the UK Strategic Nuclear Deterrent, Volume II: The Labour Government and the Polaris Programme, 1964–1970*, Abingdon, Routledge, 2017.

MacInnes, Colin, *Trident: The Only Option?* London, Brassey's, 1986.

Mallaby, Christopher, *Living the Cold War Memoirs of a British Diplomat*, Stroud, Amberley Publishing, 2017.

Moore, Richard, *Nuclear Illusion, Nuclear Reality: Britain, the United States and Nuclear Weapons 1958–64*, Basingstoke, Palgrave Macmillan, 2010.

Owen, David, *Nuclear Papers*, Liverpool, Liverpool University Press, 2009.

Salisbury, Daniel, *Secrecy, Public Relations and the British Nuclear Debate How the UK Government Learned to Talk about the Bomb, 1979–1983*, London, Routledge, 2020.

Shepherd, John, *Crisis? What Crisis? The Callaghan Government and the British 'Winter of Discontent'*, Manchester, Manchester University Press, 2015.

Simpson, John, *The Independent Nuclear State, Britain, the United States and the Military Atom*, Basingstoke, Macmillan, 1984.

Stoddart, Kristan, *Facing Down the Soviet Union: Britain, the USA, NATO and Nuclear Weapons, 1976–1983*, London, Palgrave Macmillan, 2014.

Stoddart, Kristan, *The Sword and the Shield: Britain, America, NATO and Nuclear Weapons 1970–1976*, Basingstoke, Palgrave Macmillan, 2014.

Sykes, Lynn R., *Silencing the Bomb One Scientist's Quest to Halt Nuclear Testing*, New York, Columbia University Press, 2017.

Walker, John R., *British Nuclear Weapons and the Test Ban 1954–1973 Britain. The United States, Weapons Policies and Nuclear Testing: Tensions and Contradictions*, Farnham, Ashgate, 2010.

Walker, John R., *Britain and Disarmament the UK and Nuclear, Biological and Chemical Weapons Arms Control and Programmes 1956–1975*, Farnham, Ashgate, 2012.

Walker, John R., *A History of the United Kingdom's WE177 Nuclear Weapons Programme From Conception to into Service 1959–1980*, London, BASIC, March 2019.

Zuckerman, Sir Solly, *Nuclear Illusion and Reality*, London, Viking, 1983.

Journals and Periodicals

Arms Control Today.
Bulletin of the Atomic Scientists.
International Affairs.
International Security.
Journal of Strategic Studies.
The Nonproliferation Review.
Royal Air Force Historical Society Journal.
The RUSI Journal.
Scientific American.
Survival.

Theses

Chapman, Geoffrey, *Knowledge Management and Institutional Development Within the British Nuclear Weapons Programme, 1947–1993*, PhD Thesis in the Department of War Studies, King's College London, 2020 May.

Glossary

ABM – Anti-Ballistic Missile

ACDD – Arms Control and Disarmament Department, FCO

ACNS (P) – Assistant Chief of the Naval Staff (Policy)

ACSA (N) – Assistant Chief Scientific Adviser (Nuclear), MOD

AD/DSc – Assistant Director/Defence Science

AFD – Arming and Fuzing Device

ANP – Activities Not Prohibited

APS – Assistant Private Secretary

ASW – Anti-Submarine Warfare

AUS – Assistant Under-Secretary

AWRE – Atomic Weapons Research Establishment

BNFL – British Nuclear Fuels Limited

BTWC – Biological and Toxin Weapons Convention

CCD – Conference of the Committee on Disarmament, Geneva

CD – Committee Disarmament (1979–1984), Conference on Disarmament, Geneva 1985 to present.

CNS – Chief of the Naval Staff

CPE – Chief of the Polaris Executive

CSA – Chief Scientific Adviser, MOD

CTB – Comprehensive Test Ban

CTBT – Comprehensive Nuclear Test Ban Treaty

DASO – Demonstration and Shake Down Operation

DAWP & F – Director of Atomic Weapons Production and Facilities, MOD

DCA (PN) – Deputy Chief Adviser (Projects and Nuclear), MOD

DOD – Department of Defense (US)

DOE – Department of Energy (US)

DS – Defence Secretariat

DUS (P) – Deputy Under-Secretary (Policy), MOD

ENDC – Eighteen Nation Disarmament Committee, Geneva

FCO – Foreign and Commonwealth Office

FRS – Fellow of the Royal Society

GSE – Group of Scientific Experts

HEU – Highly Enriched Uranium

IPCS – Institution of Professional Civil Servants

JCC – Joint Consultative Commission

JCS – Joint Chiefs of Staff (US)

JIC – Joint Intelligence Committee

LANL – Los Alamos National Laboratory

LLNL – Lawrence Livermore National Laboratory

MDA – 1958 Mutual Defence Agreement

MIRV – Multiple Independently Targetable Re-entry Vehicle

MOD – Ministry of Defence

NAST – Naval Air Staff Target

NPT – Non-Proliferation Treaty

NTS – Nevada Test Site

NWS – Nuclear weapons state(s)

NNWS – Non-nuclear weapons state(s)

NSC – National Security Council

NSS – National Seismic Stations

NTB – Nuclear Test Ban

PNE – Peaceful Nuclear Explosion

PNET – Peaceful Nuclear Explosions Treaty

PS – Private Secretary

PTBT – Partial Test Ban Treaty

PUS – Permanent Under-Secretary

OSI – On-site inspection

RAF – Royal Air Force

RNAD – Royal Naval Armament Depot

ROF – Royal Ordnance Factory

SACEUR – Supreme Allied Commander Europe, NATO

SALT – Strategic Arms Limitation Treaty/Talks

SLBM – Submarine Launched Ballistic Missile

SNM – Special Nuclear Material

SoS – Secretary of State

SSBN – Submarine Ballistic Nuclear

SVA – Separate Verification Agreement

TNA – The National Archives, Kew

TNWPSG – Theatre Nuclear Weapons Policy Steering Group, MOD

TTB – Threshold Test Ban

TTBT – Threshold Test Ban Treaty

VCNS – Vice Chief of the Naval Staff

UKAEA – United Kingdom Atomic Energy Authority

UKDis – UK Disarmament Delegation, Geneva

UKMis – UK Mission, Geneva

UNE – Underground Nuclear Explosion

USSR – Union of Soviet Socialist Republics

Index

For Product Safety Concerns and Information please contact our EU
representative GPSR@taylorandfrancis.com
Taylor & Francis Verlag GmbH, Kaufingerstraße 24, 80331 München, Germany

www.ingramcontent.com/pod-product-compliance
Lightning Source LLC
Chambersburg PA
CBHW060308220326
41598CB00027B/4267